Chemistry with Ultrasound

Critical Reports on Applied Chemistry Volume 28

Chemistry with Ultrasound

Edited by T. J. Mason

Published for the
SOCIETY OF CHEMICAL INDUSTRY
by
ELSEVIER APPLIED SCIENCE
LONDON and NEW YORK

° *382102x*

CHEMISTRY

ELSEVIER SCIENCE PUBLISHERS LTD
Crown House, Linton Road, Barking, Essex IG11 8JU, England

Sole Distributor in the USA and Canada
ELSEVIER SCIENCE PUBLISHING CO., INC.
655 Avenue of the Americas, New York, NY 10010, USA

WITH 23 TABLES AND 44 ILLUSTRATIONS

© 1990 The Society of Chemical Industry

British Library Cataloguing in Publication Data

Chemistry with ultrasound.
1. Chemistry. Use of ultrasonic waves
I. Mason, Timothy J. (Timothy James), *1946–* II. Society
of Chemical Industry III. Series
542

ISBN 1-85166-422-X

Library of Congress Cataloging-in-Publication Data

Chemistry with ultrasound/edited by T. J. Mason
    p.   cm.—(Critical reports on applied chemistry: v. 28)
    Includes bibliographical references.
    ISBN 1-85166-422-X
    1. Chemistry, Physical and theoretical. 2. Ultrasonic waves.
I. Mason, T. J. II. Series.
QD455.2.C48 1990
547.1'3—dc20

Photoset and printed in Northern Ireland by The Universities Press (Belfast) Ltd.

Preface

It is very seldom that the opportunity arises for a chemist to move into a new field of research involving modern technology. The drawbacks to most scientific developments are that they require expensive and often specialized apparatus and a considerable degree of expertise in a particular field of study. Added to this there is usually a restriction on the type of chemical system to which the technology applies—e.g. photochemistry requires the presence of a chromophore and electrochemistry can only be performed in a conducting medium. What makes the new topic of sonochemistry so exciting is that it has applications across almost the whole breadth of chemistry, it is quite inexpensive to get started and, with the aid of only a few simple ground rules, any chemist from undergraduate to professor can achieve exciting results in this rapidly expanding research area.

In its broadest form sonochemistry can be defined as the general study of the uses of ultrasound in chemistry. As we will see later, this definition encompasses two distinctly different uses of ultrasound, either as a source of energy capable of affecting chemical reactivity (power ultrasound) or as an analytical tool with only physical effects upon the system through which it passes (diagnostic ultrasound). It is the former type which will be the subject of this volume, for it is undoubtedly the area of most interest to the synthetic and development chemist. Power ultrasound will be reviewed in terms of its three main applications—sonochemical catalysis, organic synthesis and polymer science—with the subject of equipment and scale-up problems being addressed in the last chapter.

T. J. MASON

v

Contents

List of contributors

Robert S. Davidson Department of Chemistry, City University, Northampton Square, London EC1V 0HB, UK

Terry J. Goodwin* Biotechnology and Speciality Chemicals Centre, Harwell Laboratory, Oxon OX11 0RA, UK

James Lindley Faculty of Applied Science, Coventry Polytechnic, Priory Street, Coventry CV1 5FB, UK

John P. Lorimer Faculty of Applied Science, Coventry Polytechnic, Priory Street, Coventry CV1 5FB, UK

Timothy J. Mason Faculty of Applied Science, Coventry Polytechnic, Priory Street, Coventry CV1 5FB, UK

* Present address: British Steel Technical, Welsh Laboratories, Port Talbot, West Glamorgan SA13 2NG, UK.

1 Introduction

T. J. Mason
Coventry Polytechnic, UK

Ultrasound is not a topic which a student will normally meet on a chemistry course. It is more familiar in the context of animal communications (bat navigation and dog whistles), medical diagnosis (foetal scanning), materials testing (flaw detection) and underwater ranging (SONAR). These applications can be grouped together under

1

the heading of diagnostic ultrasound, since each relies on the same pulse-echo principle for range-finding. The first commercial application of ultrasound was precisely for this purpose as a depth gauge for ships.

The history of the generation of ultrasound dates back 100 years to the work of F. Galton, who was interested in establishing the threshold frequencies of hearing both for animals and human beings.[1] He produced a whistle with an adjustable resonance cavity which was capable of generating sound of known frequencies. With the aid of this instrument he was able to determine that the limit of human hearing is normally around 16 000 cycles s^{-1} (16 kilohertz (kHz)), although it is somewhat lower in elderly people. The whistle is an example of a transducer—a device which converts one form of energy (in this case gas motion) into another (sound). Gas-driven whistles are not used in sonochemistry because the sound intensity generated is not large enough to affect chemical reactivity. It is possible, however, to use the whistle principle with fluid motion replacing gas to produce cavitation in a liquid via a vibrating blade—this type of system will be described in detail later in this volume.

The type of transducer used in ultrasonic equipment suitable for sonochemistry converts electrical energy into sound energy. Most modern sonochemical equipment is based on the piezoelectric effect, which was discovered by the Curie brothers at the turn of the century.[2] The effect is found in some crystalline materials (e.g. quartz) and is the production of a potential difference across opposite faces of a crystal of such a material when it is subjected to a sudden compression. The inverse effect is used as the principle of modern transducers, i.e. a rapidly alternating potential when applied across opposite faces of a piezoelectric crystal will induce corresponding, alternating, dimensional changes and thereby convert electrical into vibrational (sound) energy.

Ultrasound is defined as any sound which is of a frequency beyond that to which the human ear can respond, i.e. above 16 kHz. There are, however, two distinctly different uses to which ultrasound can be put and these are identified in terms of their frequency ranges and applications:

(a) high-frequency or diagnostic ultrasound (2–10 MHz)
(b) low-frequency or power ultrasound (20–100 kHz)

Although there is a large and increasing interest in the applications of diagnostic ultrasound to analytical chemistry, such irradiation causes only temporary physical changes in the medium through which it passes and does not influence chemical reactivity. For this reason the treatment of diagnostic ultrasound will be very brief.

1.1 Diagnostic ultrasound

Each of the familiar applications of high-frequency ultrasound identified at the beginning of this chapter relies upon the release of a pulse of sound energy through a medium and the detection of an 'echo' of this sound as it returns after reflection at the surface of a solid object, a phase boundary or some other interface. Diagnostic ultrasound can be used for chemical analysis, particularly for remote sensing in flow systems.[3]

In the system shown in Fig. 1.1, the frequency of ultrasonic irradiation is fixed by the relative locations of the transducer and receiver, normally the diameter of the tube. The velocity of the sound is, however, dependent upon the medium through which it passes, so that any changes in the character of a liquid mixture—i.e. the

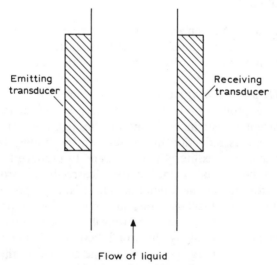

Emitting transducer

Receiving transducer

Flow of liquid

Fig. 1.1 Schematic diagram of a possible arrangement for the use of ultrasound in chemical analysis and/or monitoring.

formation of a reaction product—must result in a change in sound velocity through the medium. This will result in a difference in the time elapsed between the emission and reception of the ultrasonic pulse. With a calibration curve tied into a computer, almost instantaneous and continuous monitoring of reaction progress can be achieved.

1.2 Power ultrasound

Generally less well known than the applications of ultrasound in range-finding and related topics are the wide variety of uses which have been found for power ultrasound. A number of these are summarized in Table 1.1, most of them dating back to the 1950s or earlier. Some have fallen into disuse in recent years but, in parallel with the upsurge in interest in sonochemistry, many of these earlier applications are being given a new lease of life with recent advances in ultrasonic engineering. It is worth remembering that it was developments in plastic welding and cleaning-bath technology which gave the chemist the current tools for sonochemistry.

1.3 Cavitation

Power ultrasound influences chemical reactivity through an effect known as cavitation which was first characterized by Sir John Thornycroft and Sidney Barby at the turn of the century.[4] They were called in to investigate the poor performance of a new screw-driven destroyer (*H.M.S. Daring*) and traced the problem to an incorrect setting of the propeller blades. A wrongly set propeller does not provide efficient thrust; its rapid motion through the water produces such a tremendous reduction in pressure on the trailing faces of the blades that the water is literally 'torn apart' to produce tiny bubbles. This effect is known as cavitation. Even correctly set propellers are subject to erosion via the cavitation effect and so powerful is the cavitation that even the best brass alloys used for propellers suffer erosion damage in normal usage. As we will see, these same cavitation bubbles can be produced by the irradiation of a liquid with power ultrasound. It is the energy generated on the collapse of these bubbles which is the underlying reason for the tremendous chemical enhancements and transformations which can be achieved in sonochemistry.

Table 1.1 Some industrial uses of ultrasound

Field	Application
Biology, biochemistry	Power ultrasound is used to rupture biological cell walls in order to release the cellular contents for further studies. Originally used as the first step in the isolation of cell contents, the way is now open for use of ultrasonic disruption to release the total contents of selected cells into reaction systems and achieve enzymatic transformations without purification
Engineering	Ultrasound has been used to assist drilling, grinding and cutting. It is particularly useful for processing hard, brittle materials, e.g. glass and ceramics. Other uses of power ultrasound are welding (both plastics and metals) and metal tube drawing. There are also applications in dentistry for both the cleaning and drilling of teeth
Industrial	Pigments and solids can easily be dispersed in paint, inks and resins. Engineering articles are often cleaned and degreased by immersion in ultrasonic baths. Some less widely used applications are acoustic filtration and ultrasonic drying, atomization, precipitation, degassing and electroplating
Medicine	Although the major application of ultrasound in medicine is in ultrasonic imaging (2–10 MHz), particularly in obstetrics, power ultrasound has been used for many years in physiotherapy as an aid to massage and in particular for the treatment of muscle strains.
Plastics and polymers	The welding of thermoplastics is efficiently achieved using power ultrasound. It is also possible to initiate a radical polymerization and to degrade pre-formed polymers.

How can ultrasound produce cavitation in a liquid? Sound is transmitted through any fluid as a wave consisting of alternating compression and rarefaction cycles. We can picture the source of the ultrasonic waves (the transducer) as a piston dipping into the fluid and operating with very small but extremely rapid strokes. In this analogy, the pressure wave is clearly understood as the forward stroke into the medium which is transmitted by a series of molecular interactions through the fluid. It is the 'pull' stroke which produces the rarefaction portion of the wave. When the piston is operating at a rate of 20 000 strokes per second, ultrasound is generated in the medium. If the rarefaction wave is sufficiently powerful it can develop a negative pressure large enough to overcome the intermolecular forces binding the fluid. In this situation the molecules will be torn apart from each other to form tiny microbubbles throughout the medium. This is exactly analogous to the production of cavitation bubbles by the action of a propeller. For ultrasonic cavitation, however, there will be a compression cycle following rarefaction and this can cause the microbubbles to collapse almost instantaneously with the release of large amounts of energy. It has been estimated that temperatures of 5000 K and pressures of the order of 1000 atmospheres are generated by the collapse of cavitation bubbles generated by power ultrasound in water at 25°C.

1.3.1 Some background theory

Whenever an acoustic field is applied to a liquid, the pressure waves of the sonic vibrations create an acoustic pressure (P_a) which travels through the medium. This acoustic pressure will be applied to a system in addition to the ambient (barometric) hydrostatic pressure (P_h) which is already present in the medium. For most laboratory reactions any small additional hydrostatic pressure which might result from depth considerations can be ignored. The applied acoustic pressure is time (t) dependent and can be represented by eqn (1.1)

$$P_a = P_A \sin 2\pi f t \qquad (1.1)$$

where f is the frequency of the wave ($>16\,\text{kHz}$ for ultrasound) and P_A is the maximum pressure amplitude of the wave. By analogy with electrical vibrations it is considered that the intensity of the wave (I, the energy transmitted per second per square metre of fluid) is described as in eqn 1.2.

$$I = \frac{P_A^2}{2\rho c} \qquad (1.2)$$

where ρ is the density of the medium (kg m^{-3}) and c is the velocity of sound in that medium (m s^{-1}).

As the sound wave propagates through the liquid, it induces oscillation of the molecules about their mean rest position and thereby increases, momentarily, their mean translational energy. Although, in principle, this translational energy can be transferred *in toto,* by elastic collisions, to other molecules, and so increase their translational energy, in reality energy losses will occur owing to two effects: (a) viscous (the frictional motion of one molecule relative to another in the liquid) and (b) thermal (heat transfer from regions of high to low translational energy). It is expected therefore that the energy of the wave will be attenuated (i.e. like all sound its intensity will reduce with distance) as it passes through the medium. The extent of this attenuation can be represented by eqn 1.3,

$$I = I_0 \exp(-2al) \qquad (1.3)$$

where a is the absorption (attenuation) coefficient, I_0 is the initial intensity (energy flowing per unit area) of the sound wave, and I is the intensity at some distance l from the source. The absorption coefficient will depend not only upon the nature of the fluid and its temperature but also on the frequency of the wave (the higher the frequency the smaller the penetration. The frequency dependence is at first surprising but it is due to the fact that the total energy content of a liquid is not restricted solely to translational energy; it is the sum of many components including rotational, vibrational, molecular conformational and structural forms. It is the coupling of the sound wave with these other energy forms which increases with increased frequency and leads to the observed dependence.

As explained above, when the net negative pressure (P_c) developed in the rarefaction cycle of the sound wave is applied to the liquid $(P_c = P_a - P_h)$ such that the distance between the molecules exceeds the critical molecular distance necessary to hold the liquid intact, the liquid will break down and voids will be created i.e. cavitation bubbles will form. If we assume that the critical distance for water is 10^{-5} cm then the tensile stress, or pressure involved, can be calculated to be of the order of 10 000 atmospheres $(P_c \sim 2\sigma/R$ where $\sigma = $ surface tension).

This calculation assumes that the water is completely pure. In practice the pressure required to cause a liquid to cavitate occurs at considerably lower applied acoustic pressures than such calculations would suggest. This is due to the presence in a liquid of weak spots which lower its tensile strength. If we consider the factors which influence acoustic cavitation then we will be some way towards identifying the conditions which govern the effectiveness of the sonolysis in a chemical reacting system.

1.3.2 Factors which affect cavitation

(a) Physical properties of the solvent Each solvent will have its own particular solvent parameters so that the choice of solvent medium becomes very important when considering the reaction conditions to be used for sonochemistry. Consider the choice between water and pentane *vis-à-vis* effective cavitation. The intramolecular forces within these liquids are quite different, in that water has much stronger cohesion via hydrogen bonding than does pentane, which is held by van der Waals forces. Water also has a higher surface tension, giving water a much lower cavitation threshold than pentane (i.e. under otherwise identical conditions a greater intensity of insonation is required to cause cavitation in water than in pentane). The vapour pressure of water is considerably lower than that of pentane (see (b) below), so that it is altogether a better choice of medium for sonochemical reactions than pentane.

Naturally there will be chemical constraints upon the medium chosen in sonochemical reactions so that a solvent such as pentane— though not ideal—may have to be used. In such cases, other factors must be borne in mind.

(b) Reaction temperature In sonochemistry it is common practice to use the lowest possible reaction temperature consistent with reasonable overall reaction times. This is the direct result of the decrease in vapour pressure which accompanies the lowering of solvent temperature. From a practical point of view it is not very sensible to attempt sonochemical reactions in a solvent near its boiling point. This is because the rarefaction cycle will cause boiling of the liquid medium (as a result of the reduced pressure generated) and any cavitation bubbles formed would fill almost instantaneously with solvent vapour. Collapse of these vapour-filled bubbles during the compression cycle would be 'cushioned' thereby reducing the extremes of temperature and pressure generated. Indeed, it is quite possible that under these conditions of insonation the bubbles would not collapse at all.

(c) Irradiation frequency In order to achieve changes in chemical reactivity the sonochemist will normally use frequencies between 20 and 50 kHz. The reasons for this are twofold: firstly most commercially available equipment operates within this range, and secondly it is more difficult to achieve cavitation at higher frequencies. To understand the problems associated with producing cavitation at very high frequencies one must consider first that there will be a natural delay between applying a rarefaction wave to a fluid and the molecules of that fluid responding. At frequencies in the megahertz range, one must compensate for this 'delay' by applying a wave of greater intensity—a more powerful rarefaction—so that the effective pulling apart of molecules is greater. Unfortunately it is an almost insurmountable engineering problem to drive a very high-frequency transducer assembly at the vibrational amplitudes (intensities) required.

(d) The presence of dissolved gases When a cavitation bubble is initiated in the rarefaction cycle, it will not enclose a vacuum but will almost certainly contain some vapour of the liquid within which it is formed. It is clearly easier for that bubble to form if it is created in a solvent of high vapour pressure (see above). In the limit, this vapour pressure would be so high that the liquid would 'boil' into the bubble. The extent to which a gas is soluble in a medium depends upon the applied pressure; any sudden reduction in this pressure will release the gas (consider the effect of unscrewing the top of a shaken fizzy drink). It is therefore apparent that during the rarefaction cycle any gas dissolved in the medium would be forced (or, more explicitly, 'sucked') out of solution to form the nucleus of a cavitation bubble. From the equations for cavitational collapse (eqns (1.4) and (1.5) below), it will be seen that the ratio of specific heats of any gas in the bubble is a very important factor when considering collapse temperatures and pressures. This ratio is highest for monatomic gases (1·66) and thus sonochemistry is best conducted on a system through which a monatomic gas, usually argon, is bubbled. The gas is not very soluble in most organic media and thus is an ideal way of providing a continuous source of nucleation sites producing regular and powerful cavitation throughout the system.

This is of course the reason why ultrasound is used for degassing a liquid, since the microbubble formed in a fluid saturated in a soluble gas would contain plenty of that gas. The presence of a large amount of gas would cushion and perhaps prevent total collapse of the bubble in the succeeding compression cycle. The next rarefaction phase would then attract more gas into the bubble, and so on. At 20 000 Hz the microbubble would rapidly achieve the status of an actual bubble within the liquid. At this stage the bubble would be buoyant, float to the surface, and discharge into the atmosphere.

(e) Cleanliness of the reacting system Any particles or motes present in the solvent will act as seeds for cavitation. Such particles will have entrapped gases in their crevices and recesses which readily become the nuclei for newly forming cavitation bubbles.

(f) Reaction overpressure The cavitation threshold depends upon the rarefaction cycle generating pressures which can remove and exceed (in a negative sense) the ambient pressure on the system. Increasing the external pressure has the effect of increasing the cavitational threshold for the system by (a) requiring a larger negative pressure in the rarefaction phase and (b) decreasing the ambient vapour pressures of the reaction constituents and solvents. The net effect of an increase in overpressure is to increase the effective power of cavitational collapse—although the intensity of ultrasonic irradiation must itself be increased to achieve cavitation in the first place.

(g) Irradiation power Any increase in intensity will, in general, provide for an increase in sonochemical effect. However it must be realized that intensity cannot be increased indefinitely. With increase in the pressure amplitude (P_A) the bubble may grow so large on rarefaction that the time available for collapse is insufficient. Indeed if a sufficiently large number of such bubbles are created at the radiating face of the ultrasonic source they will provide a bubble 'cushion' which will reduce the effective coupling of the sound energy to the system.

There are thus two extremes to be considered with respect to power: (a) the threshold power which must be surpassed and (b) an optimum power beyond which further power increases have no effect.

1.3.3   Types of cavitation

It is important to recognize that the collapse of the newly-formed cavitation microbubble will not necessarily occur in the immediately succeeding compression cycle. The lifetime will depend upon whether there has been sufficient time during the growth (rarefaction) period to allow the influx of gas, or solvent vapour, into the bubble. Depending on their size, some cavitation bubbles may last for only a few cycles, giving rise to 'transient' cavitation, but others which oscillate in the acoustic field with a much longer lifetime give 'stable' cavitation. Even a stable bubble has a very short lifetime; thus, a bubble which exists for as many as 1000 cycles in a 20 kHz field has a lifetime of only 0·05 s.

Transient cavitation bubbles are voids, or vapour-filled bubbles, produced using ultrasonic intensities in excess of $10 \, W \, cm^{-2}$. They exist for one, or at most a few acoustic cycles, expanding to a radius of at least twice their initial size, before collapsing violently in the compression phase. On collapse they may disintegrate into smaller bubbles which themselves form nuclei for further cavitation. The implosion of transient cavities is very violent because there has been little time to allow gas or vapour diffusion into the bubble.

Equations (1.4) and (1.5) have been developed for the estimation of the maximum temperature (T_{max}) and pressure (P_{max}) developed within the bubble at the moment of collapse.[4] Calculations using these equations suggest that for the implosion of a bubble containing nitrogen ($\gamma = 1.33$) in water at 20°C and ambient pressure the values for T_{max} and P_{max} are 4200 K and 975 bar, respectively.

$$T_{max} = T_0 \left\{ \frac{P_m(\gamma - 1)}{P} \right\} \tag{1.4}$$

$$P_{max} = P \left\{ \frac{P_m(\gamma - 1)}{P} \right\}^{\gamma/\gamma - 1} \tag{1.5}$$

where T_0 is the ambient (experimental) temperature, γ is the ratio of specific heats of the gas (or gas/vapour) mixture , P is the pressure in the bubble at its maximum size and is usually assumed to be equal to the vapour pressure (P_v) of the liquid.

It is the existence of these very high temperatures and pressures within the collapsing transient bubble that have formed the basis for the explanation of much of the chemical reactivity induced by power ultrasound. The release of the pressure, as a shock wave through the surrounding medium, is a factor which has been used to account for polymer degradation.

Stable cavitation bubbles, because of their relative longevity, are believed to contain mainly gas and some vapour and are produced at fairly low intensities ($1-3 \, W \, cm^{-2}$). For many years it was believed

12 *T. J. Mason*

that only transient cavitation could influence chemical reactivity, but estimates of the temperatures and pressures produced in stable bubbles as they oscillate in resonance with the applied acoustic field would suggest otherwise. Although the pressures and temperatures developed in these bubbles are less than those developed on the collapse of transient bubbles, they exist for much longer and thus have more chance of affecting reactivity.

For either type of cavitation, intense local strains in the vicinity of the bubble are responsible for the many disruptive mechanical effects of ultrasound which result in some of the spectacular effects produced upon sonication of reacting systems.

1.4 The influence of ultrasound on chemical reactivity

If you were asked to identify the general methods of increasing chemical reactivity your responses would include some, if not all, of those shown in Table 1.2.

Table 1.2 General methods of increasing chemical reactivity

(i) Increase the reaction temperature
(ii) Increase the concentration of a reagent
(iii) Increase the pressure applied to the system
(iv) Use a catalyst

As a result of the copious evidence which has accumulated in the chemical literature on the effects that ultrasonic irradiation has on chemical reactivity we must now add a further option to these:

(v) Irradiate the system with power ultrasound

It has already been stated that the dramatic effects observed in sonochemistry can be ascribed mainly to the collapse of cavitation bubbles but, before we turn our attention to the many uses of ultrasound in chemistry which are reviewed in the succeeding chapters, it will be instructive first to consider the possible mechanisms by which ultrasound produces its effects.

1.4.1 Heterogeneous solid–liquid system

Without doubt it is in the field of heterogeneous chemistry that power ultrasound has had its widest applications, particularly in the area of organometallic chemistry.

Within this topic it is possible to identify two types of heterogeneous reaction which involve elemental metals:

(a) those in which the metal is a reagent and is consumed in the process;
(b) those in which the metal functions as a catalyst.

It would be tempting to explain that the ultrasonically induced enhancements in chemical reactivity which are observed in such heterogeneous reactions are due simply to the well-known cleaning action of ultrasound. It is certainly true that sonication will clean the surface of a metal and that dirty surfaces can inhibit a chemical reaction. It is because of surface contamination that many of the metals used in chemical reactions are cleaned before use; for example, copper is washed with EDTA to remove surface salts and iodine is commonly used in the preparation of a Grignard reagent to remove oxide film and promote magnesium reactivity. Sonication has something in common with both of these chemical techniques—it exposes clean or reactive surface to the reagents involved. Examination of irradiated surfaces by electron microscopy reveals 'pitting' of the surface of the metal, which acts both to expose new surface to the reagents and increase the effective surface area available for reaction. The pitting is thought to be the result of two processes resulting from cavitational collapse.[5] In the first of these, the implosion of cavitation bubbles formed from seed nuclei on the surface of the solid generates high-intensity shockwaves at the surface itself. In the second process, the collapse of a cavitation bubble in the liquid but close to the surface is asymmetric and causes a jet of solvent to impinge onto the surface, the effect is known as microstreaming.

In some cases it has been shown that the cleaning effect alone is not sufficient to explain the extent of the sonochemically enhanced reactivity. In such cases it is thought that sonication serves to sweep reactive intermediates, or products, clear of the metal surface and thus present renewed clean surface for reaction.[6] The corresponding surface sweeping effect which can be achieved by normal mechanical agitation is by no means as effective as sonication.

In heterogeneous reactions involving solids dispersed in liquids, the overall reactivity, just as with the metal surface reactions described above, will depend upon the available reactive surface area. The difference when using powders (metallic or non-metallic) is that ultrasonic cavitational collapse at, or near, the surface can lead to fragmentation and consequent reduction of particle size of the solid material, with a concomitant increase in surface area. It is important to note that the power required for such fragmentation, particularly with metal powders, is normally achieved with probe-type systems rather than baths. From a purely mechanical point of view, size reduction is an important consideration in laboratory-scale work involving ultrasonic probes, since acoustic streaming from the probe tip can often provide efficient mechanical mixing as well as particle size reduction. In contrast to this, a heterogeneous sonochemical reaction performed in an ultrasonic bath may well require additional mechanical stirring. The reason for this is that although sonication transmitted from the bath may provide sufficient power to cause particle fragmentation, it will not normally provide sufficient acoustic energy to physically stir the reaction mixture at the same time.

The efficiency of ultrasound in providing increased mass transport and surface activation in solid–liquid systems affords the opportunity to dispense with a phase-transfer catalyst (PTC). Such a catalyst is normally required to 'solubilise' the active reagent, e.g. the solid, so that it enters into the liquid phase reaction. It is necessary because without it the reaction sites are limited to the surface area of contact between the solid reagent and the liquid phase.

1.4.2 Heterogeneous liquid–liquid system

Ultrasound is known to generate extremely fine emulsions from mixtures of immiscible liquids and one of the main consequences of achieving such fine emulsions is the dramatic increase in the interfacial contact area between the liquids. This provides an increase in the region over which any reaction between species dissolved in the liquids can take place. As with the powder reactions above, this can result in the use of ultrasound in place of a PTC. However, as we will discover later, in some cases a combination of sonication and a PTC has a better overall effect than either of the two techniques alone.

1.4.3 Homogeneous liquid reaction

Some of the earliest studies in sonochemistry involved the sonolysis of water to produce H· and HO· radicals and thereby generate hydrogen peroxide. The explanation for this reactivity lies in the fact that a cavitation microbubble does not enclose a vacuum—it contains vapour from the solvent, in this case water. On collapse, these pockets of water vapour are subjected to enormous increases in both temperature and pressure. Under such extreme conditions it is not surprising that the water suffers fragmentation to generate reactive species, i.e. radicals. Some of these will be high enough in energy to fluoresce and cause the phenomenon of sonoluminescence. In organic solvents, many examples have been reported of radical formation during sonication.

The shock wave produced by bubble collapse, and indeed the propagating ultrasonic wave itself, are also thought to influence chemical reactions by the disruption of any structure (e.g. hydrogen-bonding) which the solvent itself might possess. Such solvent disruption could also influence reactivity within the system by altering solvation of the reactive species present. Calculations show that the shock wave associated with cavitational collapse is sufficiently energetic to cause the degradation of polymeric species.

1.5 Illustrative examples of the advantages of using ultrasound to enhance chemical reactivity

(1) Rate acceleration in a homogeneous solvolysis reaction [7]

The extent to which ultrasound can affect the rate of solvolysis of 2-chloro-2-methylpropane in aqueous ethanolic mixtures depends upon a number of factors, including solvent composition and reaction temperature (eqn (1.6)). As with most sonochemical reactions, the effect of sonication (compared to a control reaction under otherwise similar conditions) is greater as the temperature is reduced. For this solvolysis, enhancement was also found to increase with increased solvent structure and this was equated with the ability of ultrasound to disrupt intermolecular hydrogen bonding. Combining both of these factors, the reaction showed a 20-fold rate increase at 10°C in 50% ethanol.

$$(CH_3)_3C\!-\!Cl + H_2O \rightleftharpoons (CH_3)_3C^+Cl^- \rightarrow (CH_3)_3C\!-\!OH + HCl \quad (1.6)$$

(2) Avoidance of forcing conditions [8]

Classical methodology for the production of transition metal carbonyl anions from metal halide, sodium and tetrahydrofuran requires high pressure (200 atmospheres) and an elevated temperature (160°C). With ultrasound these conditions can be dramatically reduced to run in a vessel requiring only 4.4 atmospheres and a much lower temperature (10°C) (eqn (1.7))

$$VCl_3\cdot(THF)_3 \xrightarrow{\text{CO/Na/THF}} V(CO)_6^- \quad (1.7)$$

(3) Simplification of procedures—safer methodologies [9]

Finely divided, highly reactive 'Rieke' metal powders are normally prepared by reducing metal halides with potassium metal in refluxing tetrahydrofuran. Powders of the same activity can be obtained via a much less hazardous procedure using lithium metal in place of potassium in THF at room temperature immersed in an ultrasonic bath (eqn (1.8)). Thus the traditional method of producing Rieke copper powder, which requires an 8 h reflux of the metal halide with potassium in THF can be reduced to less than 40 min by the sonochemical route.

$$MX_n + nA \rightarrow M^* + nAX$$
$$MX = \text{metal halide; } A = Li, Na, K \quad (1.8)$$

(4) Particle size reduction and continuous surface activation [6]

The Ullmann coupling reaction of 2-iodonitrobenzene using commercial copper bronze at 60°C in dimethylformamide shows a 64-fold rate increase when subjected to sonication (eqn (1.9)). Experiment has shown that this coupling reaction, when performed conventionally using previously sonicated copper bronze, is also subject to rate enhancement, but only to a small extent. This latter result is explained entirely by the reduction in particle size of the metal powder used. The very large enhancement achieved when sonication was applied continuously throughout the reaction must then be due to continuous surface activation of the metal. Two additional benefits accruing from sonication of this reaction are that (a) a far smaller excess of copper is required and (b) the metal powder does not adhere to the side walls of the reaction vessel.

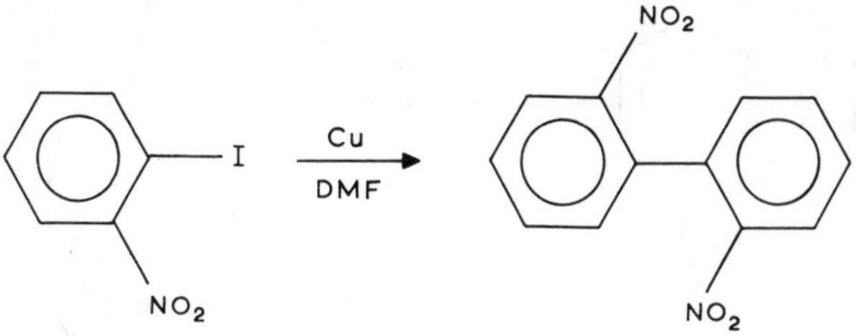

$$(1.9)$$

(5) Avoidance of phase-transfer catalysts [10]

This has been referred to above both in terms of solid–liquid and liquid–liquid systems. A very good example is the cyclopropanation of styrene, in which a 96% yield can be achieved in 1 h without the need for a phase-transfer catalyst (eqn (1.10)). The same reaction run as a control with mechanical stirring afforded a yield of only 30% after the extended reaction time of 16 h.

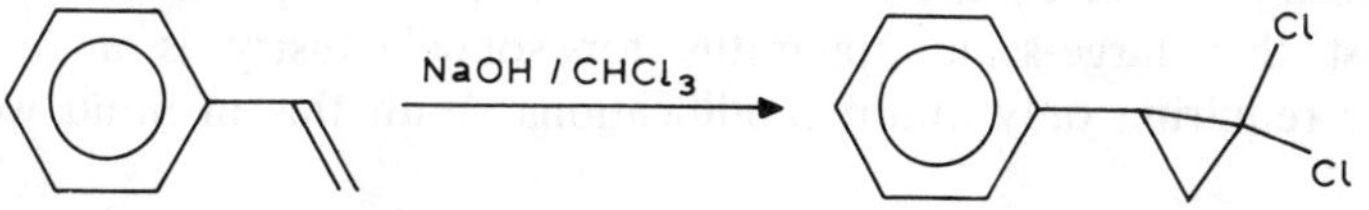

$$(1.10)$$

(6) Change in reaction pathway [11]

Conventional stirring of alumina and KCN in a mixture of benzyl bromide and toluene affords the Friedel–Crafts product (alumina acting as the catalyst). Under sonication this reaction follows a different pathway and benzyl cyanide is produced as a result of the direct displacement of Br by CN^- (eqn (1.11)). The explanation

offered for this change is that sonication impregnates the alumina
surface with CN^- ion at its Lewis acid sites, thus masking them so that
the reaction proceeds via nucleophilic displacement.

(1.11)

1.6 The uses of ultrasound in chemical technology

For many years, power ultrasound has been used in chemical
processing for a variety of purposes. Some of these techniques are
analogous to the types of application that might be found in
sonochemistry. Certainly the hardware used in processing would tend
to suggest that large-scale apparatus for sonochemistry is a real
possibility requiring only minor modifications. With this in mind we

Table 1.3 Some processing techniques
which have been improved by the application
of ultrasound

  (i) Atomization
 (ii) Precipitation and crystallization
(iii) Filtration
 (iv) Membrane permeability
  (v) Electrochemistry
 (vi) Powder dispersal and mixing

will examine a few of the methods of ultrasonic processing and related topics which have been reported in the patent and general literature (Table 1.3). It is perhaps important to note that several of these techniques were tried out many years ago and, since then, some uses have been discontinued.[12,13] It is only with the resurgence of interest in sonochemistry and the resulting interest in ultrasonic engineering that these older processes are being resuscitated.

1.6.1 Atomization

Conventional atomizers achieve their effects by forcing a liquid at high velocity through a small aperture. This process is clearly subject to the limitation that the material to be atomized must be in the form of a fairly free-flowing liquid in order that the required velocity of ejection can be attained. Even if this condition is met there is another serious drawback to the continuous running of conventional atomizers, namely that the small orifices which must be employed are subject to blockage. Atomization via ultrasound provides solutions to these problems because the flow passages are larger and can therefore be used for viscous materials and slurries. In addition, the ultrasonic vibrations employed for atomization provide the additional benefit of making the orifices self-cleaning.

There are three basic types of ultrasonic atomizer, one which is gas driven (whistle type) and two which are driven electrically either at high frequency (MHz range) or low frequency (kilohertz range, horn type).

(a) Gas driven atomizers are mostly based on the Hartmann whistle, in which pressurized gas, normally air, is directed into a resonant cavity which is the source of ultrasonic energy. The size of the cavity determines the frequency of the whistle. The liquid to be atomized is introduced into this cavity, where it is shattered into very fine droplets. In the Lucas Dawe Sonimist system, air by-passing the resonator carries the atomized droplets downstream as a soft, low-velocity spray which emerges from an orifice which is large enough to eliminate blockage problems. This type of atomizer can be designed to handle flow rates in the range 1–5000 litres/h. The major applications are in humidifying systems, dust control, gas cooling and coating.

(b) Electrically driven high-frequency atomizers are somewhat restricted in industrial usage because of their small volume-handling capability. This restriction is simply because transducers which operate in the megahertz range are themselves tiny. Nevertheless, with the capability of delivering droplets in the 1–5 μm range, 'nebulizers' employing such transducers have found applications where small throughputs are required, such as in medical inhalation therapy. A common method of operation involves direct atomization of the liquid on the face of the transducer. The tiny droplets produced need no propellant gas and simple inhalation enables them to find their way into some of the tiniest airways of the lung.[14]

(c) Electrically driven low-frequency atomizers are capable of handling large volumes and offer two methods of atomization. In the first, the material is passed through the centre of the sonic horn and is emitted at the vibrating tip. Although this type is normally restricted to usage with fairly low viscosity liquids, it is less subject to jet blockage since the 'nozzle' is under constant agitation. The second design is similar to that described above for high-frequency models. The material to be atomized is passed through an orifice (not restricted in size) onto a plate, vibrating at ultrasonic frequency, causing atomization. A great virtue of this design is that it can be used for very viscous fluids because the orifice size is not restricted. For the same reason there will be fewer blockages.

Ultrasonic atomization offers several advantages over traditional methodologies.

(i) Ultrasonically atomized sprays are not produced via a high-velocity jet, so that particle velocities are much smaller. When used for coating, impact adhesion is therefore much less of a problem—there will be less 'bounce back' from the source.
(ii) The particle size of sonically atomized sprays can be accurately controlled by variation in either ultrasonic power or frequency.
(iii) Ultrasonic atomization can be used to produce particulate sprays from such 'hostile' liquids as molten glass and metal. In such cases the particles are spherical and the particle size distribution is narrow.[15]

1.6.2 *Ultrasonic crystallization and precipitation*

Power ultrasound, by virtue of producing cavitation bubbles and its subsequent effect on particle fragmentation, has proved to be extremely efficient as a method for forcing solids out of supersaturated solutions. It will produce extremely finely divided and uniform particles since the 'seeds' as they form and grow are themselves broken down and dispersed to produce further seeding throughout the medium.

An application of this technology is in the preparation of dispersions of an insoluble drug in a liquid medium for oral or subcutaneous administration. In such cases, the extremely small particle size both gives a stable suspension of the drug and permits more rapid assimilation into the body. An ultrasonic method of producing procaine penicillin can be found in the patent literature.[16] Separate solutions of a procaine and a penicillin salt are mixed in an ultrasonic reactor to yield a product with a particle size which is smaller and much more uniform (5–15 μm) than that produced conventionally (10–200 μm).[16]

An additional industrial benefit of sonochemical precipitation is that sonication prevents encrustation of precipitated solid on any cooling coil used in the cooling process. As a result there will be a more evenly distributed overall cooling rate in such systems.[17]

1.6.3 Ultrasonically assisted filtration

In the 1970s a number of reports of the advantages of acoustic filtration were published. A review of these methods together with some original work reports flow rate improvements of between 10- and 300-fold using traditional filter membranes of metallic mesh or sandstone.[18] Power ultrasound may be applied either to the mixture above the filter or to the filter itself and there would appear to be an advantage in introducing the energy in a 'pulsed' mode.

There are essentially two effects of ultrasonic irradiation which can improve filtration: (a) the agglomeration of finely particulate matter in the 'cake' and (b) the provision of sufficient vibrational energy to the cake to keep particles partly suspended, thus providing more 'channels' for solvent elution. A particularly spectacular case is that of coal slurry. It is remarkably difficult to remove water from such slurries by conventional filtration techniques; indeed, vacuum filtration will only reduce 60% aqueous slurry to 40%. Unfortunately, coal slurry is not combustible above 30% moisture content, but ultrasonic-assisted vacuum filtration can be used to produce a combustible product directly (20% moisture content).[19]

All conventional filtration techniques suffer from the need for a membrane or filter pad to separate the solids from the liquid phase. Filters are notorious for clogging, and, indeed, the finer the particles, the slower the filtration rate and the more likely the system is to clog. Clearly, it would be a great advance in separation technology if the membrane itself could be eliminated. This approach has met with some success with the aid of ultrasonic standing waves. When two transducers are placed in a liquid medium, separated by a distance corresponding to an integral number of wavelengths of the sound wave in that medium, standing waves will be produced when the transducer frequencies are matched. In this situation any tiny particulate contaminants in the liquid are found to collect in regions at half-wavelength distances on the axis of the ultrasonic beam. If the transducers are energized at slightly different frequencies, the standing waves migrate slowly towards one of the transducers and carry the particles with them. Thus, the particles tend to concentrate at that transducer. If the liquid between the transducers is flowing and the flow is split beyond the ultrasonic field one half of the flow will be enriched in particulate matter and the other depleted. If the split flows are now recycled, it is possible to attain a much greater efficiency of separation.[20]

Since the mobility of particles in the ultrasonic field is related to their size and shape, there exists an interesting extension of this work. It might be possible to perform a 'mass-spectrometric' type of separation of particles in terms of particle dimensions.

1.6.4   Ultrasound-mediated membrane permeability

For many years the heating effect of power ultrasound applied externally to the body has been used in physiotherapy as an aid to massage for the treatment of various types of muscular strains. This type of ultrasonic massage can also be used to enhance the absorption of medicaments through the skin. In the particular case of the external application of hydrocortisone to swine, the uptake of the drug into muscle tissue has been shown to be increased by 300% using ultrasonic massage.[21]

Attempts have been made to use ultrasound to modify polymer membrane permeability *in vitro*. One aim is to develop a method of ultrasonically triggering the release of a drug encapsulated in a synthetic membrane implanted in the body. In one such model investigation it was found possible to control the transport of NaCl across a nylon membrane which had previously been coated in a quaternary ammonium compound. Normally, the coating on the nylon restricts NaCl transport, but insonation causes the coating to change into a liquid crystalline state and this permits easier transport of the NaCl.[22] Ultrasound is also capable of increasing the permeability coefficient of hydrocortisone in cellulose by 23% in an aqueous solution at 25°C.[23]

1.6.5 *Ultrasound in electrochemistry*

From a knowledge of the effects of ultrasound on a metallic surface one might expect to find at least three advantages which could accrue if sonication were used as an adjunct to electrochemical processing.

(a) The evolution of any gas at an electrode surface during electrolysis interferes mechanically with the passage of the current. If an electrolysis of this type is sonicated, then every generation of a bubble at an electrode surface provides a nucleus for cavitation. The bubbles will therefore be displaced from the surface as they are formed and vented to the atmosphere as a result of the known degassing effect of ultrasonic irradiation. Thus, sonication of an electrolysis would be expected to prevent gas bubble accumulation at the electrode.

(b) Cavitational collapse at the electrode surface causes disruption and agitation of the electrode double layer. This reduces ion depletion in the region and assists ion transport for the duration of an electrochemical process.

(c) Sonication will serve to clean and activate an electrode surface. This prevents the electrode from fouling and is particularly beneficial in electroplating.

It is perhaps in the field of electroplating that ultrasound has found its greatest use. There are two simple methods of sonicating an electroplating system. One can irradiate either the whole plating bath or simply one of the electrodes, and both methods have proved beneficial. Nickel plating can be improved by sonication of the plating bath, giving an increased deposition rate and an increase in plating current.[24] This latter effect is significant because the plating current normally falls during the process owing to polarization. Chromium plating can be improved when ultrasound is applied directly to the cathode.[25] The resulting plating shows increased microhardness (10%), increased microcrack production and better brightness compared with that produced with conventional methodology.

An area of considerable interest over recent years has been the production of conducting polymer films by electrodeposition. Using the beneficial effects of ultrasound, great improvements in the quality of electrochemically produced polythiophene have been achieved.[26] Films deposited conventionally gradually became brittle as the electrolytic current density exceeds $5\,\mathrm{mA\,cm^{-2}}$, whereas flexible and tough films are obtained even at high current density when the electrolytic solution is subjected to irradiation with ultrasound.

1.6.6 Ultrasonic dispersal and mixing

Ultrasound can be used to deagglomerate material and to reduce particle size (see above). Another mechanical effect of power ultrasound is that it can cause extremely efficient mixing, not only of immiscible liquids to form emulsions, but also to disperse powders into liquids. This has been scaled up for use in industrial processing.

The production of polymeric resins involves heating acids and glycols to a high temperature then adding styrene as the temperature is reduced. It is the styrene which takes part in curing when the activator is added. In a process developed by Scott Bader at Wellingborough in the UK, the styrene is added and mixed with a traditional stirrer. After this stage some pyrogenic silica is stirred into the resin (the quantity depending upon the desired final specification of the resin). In order to obtain product uniformity it is now necessary to produce a completely even dispersion of the silica throughout the medium. This is achieved by passing the pre-mix through an ultrasonic homogenizer which is of the whistle type and capable of handling 12 000 litres/h. This disperses the solid as a fine powder throughout the resin. The procedure ensures the correct thixotropic characteristics of each type of resin.

Using this process, Scott Bader has been able to reduce the amount of pyrogenic silica required by some 40% compared with conventional methodology. This obviously makes a considerable saving of a very expensive ingredient. Pigments can also be dispersed into the resin at the sonication stage to produce the gel coat used as the final layer in the building up of glass-fibre components.

1.7 References

1. F. Galton, *Inquiries into Human Faculty and Development*. MacMillan, London, 1883.
2. J. Curie & P. Curie, *Compt. Rend.*, **91**, 294 (1880); *idem.*, *ibid.*, **93**, 1137 (1881).
3. R. C. Asher, *Ultrasonics*, **27**, 17 (1987).
4. B. E. Noltingk & E. A. Neppiras, *Proc. Phys. Soc. B (Lond.)*, **63B**, 674 (1950); *idem.*, *ibid.*, **64B**, 1032 (1951).
5. T. J. Mason & J. Lindley, *Chem. Rev.*, **16**, 275 (1987).
6. J. Lindley, P. J. Lorimer & T. J. Mason, *Ultrasonics*, **24**, 292 (1986).
7. T. J. Mason, J. P. Lorimer & B. P. Mistry, *Tetrahedron*, **26**, 5201 (1985).
8. K. S. Suslick & R. E. Johnson, *J. Am. Chem. Soc.*, **106**, 6856 (1984).
9. P. Boudjouk, D. P. Thompson, W. H. Ohrbom & B. H. Han, *Organometallics*, **5**, 1257 (1986).
10. O. Repic & S. Vogt, *Tetrahedron Lett.*, **23**, 2729 (1982).
11. T. Ando, S. Sumi, T. Kawate, J. Ichihara & T. Hanafusa, *J. Chem. Soc., Chem. Commun.*, 439 (1984).
12. P. K. Chendke & H. S. Fogler, *Ultrasonics*, 31 (1975).
13. M. N. Topp & P. Eisenklam, *Ultrasonics*, 127 (1972).
14. P. Herzog, P. O. Norlander & C. G. Engstrom, *Acta Anaesth. Scand.*, **8**, 79 (1964).
15. R. Pohlman, K. Heisler & M. Cichos, *Ultrasonics*, 11 (1974).
16. R. R. Umbdenstock, US Patent, 2 727 892 (1955).
17. J. K. Skrebowski & J. Williamson, British Patent, 1 159 670 (1969).
18. L. Bjorno, S. Gram & P. R. Steenstrup, *Ultrasonics*, 103 (1978).
19. H. S. Muralidhara, B. K. Parekh & N. Senapati, US Patent 4 747 920 (1988).
20. J. M. Hutchinson & R. S. Sayles, *Ultrasonics International 87 Conference Proceedings*, Butterworths, London 1987, pp. 302.
21. J. E. Griffin & J. C. Touchstone, *Am. J. Phys. Med.*, **42**, 77 (1963); *idem. ibid*, **51**, 62 (1972).
22. Y. Okahata & H. Noguchi, *Chem. Lett.*, 1517 (1983).
23. T. N. Julian & G. M. Zentner, *J. Pharm. Pharmacol.*, **38**, 871 (1986).
24. S. R. Rich, *Proc. Am. Electroplaters' Soc.*, **42**, 137 (1955).
25. E. Namgoong & J. S. Chun, *Thin Solid Films*, **120**, 153 (1984).
26. S. Osawa, M. Ito, K. Tanaka & J. Kuwano, *Synthetic Metals*, **18**, 145 (1987).

2 Sonochemical aspects of inorganic and organometallic chemistry including catalysis

J. Lindley
Coventry Polytechnic, UK

2.1 Introduction

Acoustic waves are mechanical oscillations propagated through an elastic medium. In a fluid there is a secondary time-independent motion, called acoustic streaming, which arises from non-linear coupling of the first-order vibrations and gives rise to the well-known stirring action of ultrasound. Acoustic streaming leads to improved thermal mixing and mass transport within gases and liquids and also has the ability to reduce diffusion layers between interfaces; consequently it will be expected to have a marked effect on processes which

27

are diffusion controlled. At sufficiently high intensities, ultrasound produces cavitation in liquids. Cavitation is due to the formation and growth of microbubbles during the rarefaction phase of the acoustic wave and their subsequent violent collapse during the compression cycle of the wave. There are two important effects of cavitation. Firstly, as a result of adiabatic compression the cavitation bubble contents are heated to temperatures of up to 5000 K. Secondly, the implosion of cavitation bubbles produces high-energy shock waves with pressures of several thousand atmospheres. Under these conditions excited molecules and radicals may be produced and reactivity is enhanced by the increased number of molecular collisions.[1]

In this chapter the effects of ultrasound on three seemingly diverse topics are reviewed, although it will soon become clear that it is in fact simply the same set of ultrasonic phenomena which are being exploited in a variety of ways.

2.2 Metals

There is a long history of the use of high-intensity ultrasound in metals technology. Areas of application include crystallization of melts,[2] metal forming,[3] metal finishing,[4] the production of metal powders,[5] metal joining[6] (soldering, welding). As these topics are in the domain of metallurgical technology they will be considered only in outline and the interested reader is referred to reviews[6] on this area.

2.2.1 Crystallization of metals

The application of ultrasound to a metal melt generally leads to metals with improved grain refinement and homogeneity; see Figs 2.1, 2.2 and 2.3.[7] This improvement in grain refinement also leads to an improvement in the mechanical properties of the metals such as tensile strength, impact strength and hardness (Table 2.1). The acoustic fields also lead to degassing of metals and to a more even distribution of insoluble impurities in the metals.

Ultrasonic waves at intensities below the cavitation threshold have little effect on the rate of nucleation or the degree of supercooling (metastability threshold) necessary for normal crystallization, although the sonic field gives rise to acoustic streaming, which in turn leads to

Fig. 2.1 Macrostructure of an aluminium alloy (a) control (b) ultrasonically treated. (Reproduced with permission from O. V. Abramov and V. I. Teumin, in *Physical Principles of Ultrasonic Technology* Vol. 2, ed. L. D. Rozenberg, Plenum Press, 1973, p. 145.)

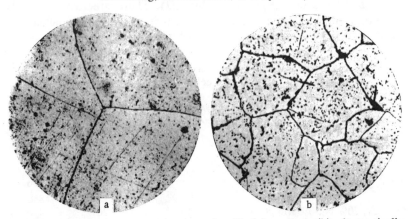

Fig. 2.2 Microstructure of chrome steel × 65: (a) control, (b) ultrasonically treated. (Reproduced with permission from O. V. Abramov and V. I. Teumin, in *Physical Principles of Ultrasonic Technology* Vol. 2, ed. L. D. Rozenberg, Plenum Press, 1973, p. 145.)

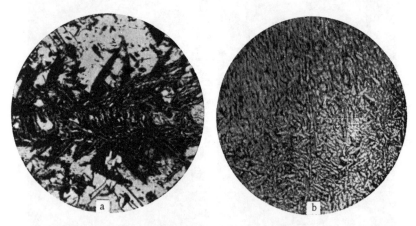

Fig. 2.3 Microstructure of brass × 65: (a) control, (b) ultrasonically treated. (Reproduced with permission from O. V. Abramov and V. I. Teumin, in *Physical Principles of Ultrasonic Technology* Vol. 2, ed. L. D. Rozenberg, Plenum Press, 1973, p. 145.)

improved mass and thermal transport within the metal and as a consequence will give rise to improved homogeneity in the metal ingot.[8] However, in the presence of ultrasonic fields above the cavitation threshold for the melt, a substantial increase in both the rate of nucleation and a decrease in the degree of supercooling required for crystallization is observed (see Table 2.2). There are two proposals concerning the role of cavitation in improving the rate of nucleation. Firstly, during the expansion phase of the cavitation bubble, liquid

Table 2.1 Effect of ultrasound during crystallization on the tensile strength of steels

Sample	Crystallization condition[a]	Test temperature	Tensile strength (MPa)
0·5% C steel	C	20	490
	US	20	630
1% C steel	C	20	489
	US	20	850
20% Cr, 20% Ni, 3% Mo steel	C	20	490
	US	20	550

Source: Abramov, O. V. (1987). *Ultrasonics,* **25,** 73.
[a] C, no ultrasound; US, in an ultrasonic field.

Table 2.2 Effect of ultrasound on the metastability threshold and nucleation rate of metals

Substance	Crystallization conditions[a]	Metastability threshold $\Delta T / °C$	Minimum time for crystallization centre to develop (*min*)
Bismuth	C	12·0	0·20
	US	2·0	0·05
Antimony	C	25·0	10·0
	US	2·0	0·05

Source: Abramov, O. V. (1987). *Ultrasonics*, **25**, 73.
[a] C, no ultrasound; US, in an ultrasonic field.

evaporates into the bubble which causes a localized fall in temperature around the bubble. This in turn leads to an increase in the degree of supercooling and nuclei are formed which subsequently become dispersed by the shock wave produced during the collapse phase of the bubble. Secondly, the increase in pressure within the system due to the shock waves resulting from bubble collapse effectively raises the melting point of the medium and is analogous to an increase in supercooling within the medium and can give rise to an increase in the number of nucleation sites.

Generally, cavitation occurs more easily in liquids which contain solid impurities; these impurities contain crevices which trap small amounts of gas, which during the rarefaction cycle of the acoustic wave are sucked out and can serve as nuclei for the formation of cavitation bubbles. The addition of small amounts of non-metallic compounds to metal melts leads to substantial reductions in the cavitation threshold. This reduction in cavitation threshold gives significant advantages in the grain refinement of metals. Thus, insonation of pure tin at 12–15 kHz, 120 W gives a 20-fold decrease in grain size, whereas tin containing 0·5% silica shows a 50–70-fold decrease in grain size at only 90 W.[7] In addition to lowering the cavitation threshold, these impurities become metallized and serve as nuclei for the crystallization process.

The application of ultrasound to metal melts has also been shown to be beneficial during the continuous casting of metals, when, in addition to improvements in homogeneity and grain structure, there are significant improvements in surface finish and the ease of processing.[9]

2.2.2   Effects of ultrasound on solid metals

Ultrasound, when applied to stressed metals, has been observed by many workers to lead to a reduction in mechanical properties such as tensile strength. At vibrational amplitudes below a certain level, the tensile stress returns to its original value on removal of the ultrasound. This type of work softening is known as the Blaha effect (Fig. 2.4).[10] At vibrational amplitudes above this level, the tensile stress returns to a higher level than the original on removal of the ultrasound—i.e. the metal is work hardened. Both work hardening and work softening are proportional to the intensity of the ultrasonic field but are independent of frequency in the range 15 kHz–1·5 MHz. The mechanism of these effects has been a source of controversy. There appear to be three major contributions. Firstly, in some cases the effects are elastic in origin and arise from a simple stress superposition. Secondly, a reduction in tensile stress may arise as a result of a decrease in the effective Young's modulus of the metal as a result of non-linear elasticity associated with the motion of dislocations and/or a rise in temperature of the stressed sample.[11] Thirdly, a reduction in tensile stress can arise from an increase in the plastic strain, caused by interaction of ultrasonic vibrations with dislocations. It is well known that application of stresses of sufficient amplitude will cause dislocation motion and multiplication which leads to permanent plastic deforma-

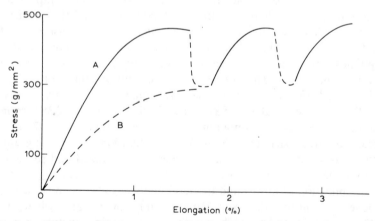

Fig. 2.4   Effect of 800 kHz ultrasound on tensile deformation of zinc crystals. A, intermittent application of ultrasound B, continuous application of ultrasound.[10]

tion. This interaction between ultrasonic waves and dislocations can lead to work softening and tensile stress reduction by increasing the amount of plastic strain and work hardening by causing the dislocations to move more slowly owing to interactions with themselves and with lattice defects.[12]

The absorption of ultrasonic energy by metals is proportional to the square of the vibrational amplitude.[13] A plot of energy absorbed against stress amplitude (Fig. 2.5) tends to a limiting value of the stress which corresponds to the fatigue limit of the metal; for example, 155 MPa for brass.[6] The increase in temperature of metals at low amplitude (i.e. intensity of ultrasound) is quite small. However, insonation at amplitudes greater than the fatigue limit can cause temperature rises of 500–800°C.

Metals immersed in cavitating liquids are subjected to high-energy shock waves which impinge on the surface of the metal, thereby causing considerable surface damage and erosion. This phenomenon

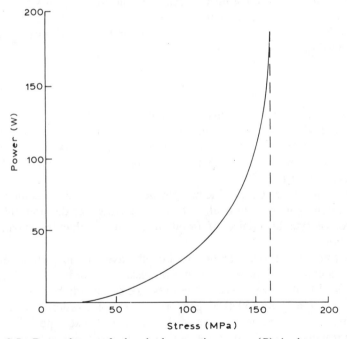

Fig. 2.5 Dependence of absorbed acoustic power (P) in brass on stress amplitude (σ).

has been exploited in commercial cleaning operations,[14] and is responsible for the activation of metals used in synthesis and catalysis. These topics are discussed in Sections 2.4 and 2.5.

The application of high-intensity ultrasound to alloys has been reported to give large increases in the rate of precipitation hardening. Thus, 3% copper–aluminium alloy when treated with ultrasound (22 kHz, 125 MPa) showed a 60-fold decrease in the time required to reach maximum hardness over a control sample. Additionally, the insonated sample shows an increase in dislocation intensity and a reduction in size of the precipitates.[15]

2.2.3 Production of metal powders

Ultrasonic vibrations below the surface of a liquid produce capillary waves on the free surface of the liquid. When the amplitude of vibration exceeds a threshold value, the capillary waves become unstable and throw off a mist of droplets, the majority of which have diameters equal to one-quarter of the capillary wavelength. Thus, droplet diameter (d) in the case of a liquid with a free surface in air or vacuum is given by eqn (2.1):

$$d = \frac{1}{2}\left(\frac{\sigma\pi}{\rho f^2}\right) \tag{2.1}$$

where f = frequency, σ = surface tension and ρ = density of the liquid. This method of ultrasonic atomization is widely used in the production of metal powders. In one method, molten metal is allowed to fall onto a vibrating plate, which produces a fine mist of particles which solidify in a cold gas stream.[16]

The advantage of producing powders in this way include the production of more spherical particles, a narrow size distribution, the control of size by choice of frequency, and a reduction in surface oxidation.

Ultrasound has been shown to be effective in the production of active metal powders for use in synthesis. The reduction of metal halides with lithium in THF in the presence of low-intensity ultrasonic fields (cleaning bath, 50 kHz) gives rise to metal powders which have reactivities comparable to those of the so-called Rieke type,[17,18] which are normally prepared by an experimentally more difficult procedure involving the reduction of the metal halide with potassium. Thus,

powders of Zn, Mg, Cr, Cu, Ni, Pd, Co and Pb were obtained in less than 40 min by this procedure, compared with reaction times of 8 h using the Rieke method. Control reductions of the metal halides with lithium in tetrahydrofuran (THF), which were mechanically stirred, took up 26 h for completion. In cases where the metal halide is unstable in THF, the addition of an electron-transfer catalyst such as naphthalene is recommended. These metal powders show enhanced reactivity in organic synthesis such as the Reformatsky and Ullmann coupling reactions.

A more active form of magnesium powder is formed when magnesium, in the presence of anthracene, is subjected to low-intensity ultrasound. The magnesium produced in this way is an excellent reducing agent for metal salts.[19] Insonation of alkali metals is a simple and convenient method for the production of colloidal metals. The method is sensitive to the solvent vapour pressure, which has a marked influence on the energetics of cavitation. Thus, colloidal potassium can be prepared in a few minutes in toluene. However, sodium requires a solvent of lower vapour pressure such as xylene, which allows more powerful cavitation to occur.[20] These alkali-metal dispersions are particularly useful reagents in organic reactions such as the Dieckmann and Wittig reactions. Highly dispersed mercury emulsions are also conveniently prepared by insonation with low-intensity ultrasound.[21] Insonation (20 kHz) of copper bronze powder in dimethylformamide leads to a rapid reduction in average particle size from 90 μm to 20 μm and gives a narrower size distribution.[22]

2.2.4   Electrodeposition of metals

The electroplating and electroless plating of a wide range of metals and alloys from electrolyte solutions in the presence of ultrasonic fields has been found to produce deposits with improved physical and mechanical properties, and has enabled plating to occur at higher current densities and hence higher plating rates. There are several comprehensive reviews on this topic.[23-27] The intense agitation caused by both acoustic streaming and cavitation lead to a reduction in polarization at electrode surfaces and to a reduction in diffusion-layer thickness. In a study of the electrodeposition of copper, Drake,[28] estimated that the diffusion-layer thickness was reduced from 200 μm with no stirring to between 20 and 30 μm (depending on angle of inclination of the ultrasound source) in ultrasonic fields of 1·2 MHz and intensity of 4 W cm^{-2}. Under these conditions only acoustic streaming is possible. However, in the presence of ultrasound of 20 kHz frequency and intensity 4 W cm^{-2} the diffusion-layer thickness is further reduced to 3·4 μm and the limiting current density is increased 50-fold. The reduction in polarization in ultrasonic fields should lead to an increased grain size in the deposit. However, in many cases ultrasound leads to a reduction in grain size. It is suggested that the relative change in degree of passivity of the electrode and the change in concentration polarization determine the grain size of the deposit. When the degree of passivation is higher than the decrease in concentration polarization, small crystals should be formed; when the degree of passivation is less than the change in concentration polarization, large crystals should be formed.[29] The decrease in grain size of ultrasonically deposited metals has been cited as the principal cause of increased microhardness of deposits which is observed for a wide range of electrodeposited metals.[29,30] However, some workers consider that this increased microhardness is a direct result of work hardening of the metal surface owing to bombardment of the surface by high-energy shockwaves produced by collapsing cavitation bubbles.[31] Electrodeposits of metals produced in ultrasonic fields often have superior brightness to those produced normally; for example, the reflectivity of nickel deposited under ultrasonic conditions is up to 300% greater than that of control samples.[29] This increase in brightness is also attributed to the effects of cavitation.[32] The cavitation shockwaves cause cavitation erosion of the growing surface, which has the effect of inhibiting growth perpendicular to the surface and results in a much smoother deposit.

Metal coatings which are electrodeposited in the presence of ultrasonic fields generally have lower internal stress. Internal stress may be compressive or tensile; generally tensile stresses should be avoided since they may lead to the deposit breaking from the substrate. Reductions in internal stress of up to 50% have been reported for copper deposited in an ultrasonic bath, compared with a stirred control.[26]

A common cause of internal stress is the incorporation of atomic hydrogen into the substrate or deposit: so-called hydrogen embrittlement. The production of gaseous hydrogen at the cathode surface is facilitated by ultrasound, which lowers the potential at which the gas is liberated and assists in its desorption, thereby decreasing the tendency for atomic hydrogen to be incorporated into the metal. This desorption of gases by ultrasound also leads to deposits with reduced porosity.

Ultrasound is also beneficial during the electrodeposition of alloys, where in addition to the above-mentioned effects there is quite often a change in the composition of the alloy. For example, brass electrodeposited under stirred conditions contained 85·4% Cu and 13·9% Zn at low current density, whereas the copper content of brass deposited in an ultrasonic field (18·5 kHz, 0·8 W cm^{-2}) fell to 62·6%.[33] Similar observations have been reported for the electrodeposition of nickel–iron alloys.[34]

2.3 Inorganic compounds

2.3.1 Crystallization, dissolution and intercalation of metal compounds

As long ago as 1927, Loomis and Richards reported the acceleration of the rate of crystallization of supersaturated sodium thiosulphate solutions in the presence of ultrasound.[35] Since then there has been a considerable increase in activity in this area. Basically, there are two effects. Firstly, with ultrasonic fields at cavitation levels increases occur in both the rate of formation and the rate of dispersion of crystallization centres (nuclei). This often results in crystals with smaller particle size and narrower particle-size distribution. The crystal habit is often different in crystals produced in acoustic fields as they tend to show less dendritic growth, which gives rise to crystals with a more equiaxed form. Secondly, with ultrasonic fields below the cavitation level, acoustic streaming is effective in increasing mass-transport of crystal-building material to the growing cyrstals, which can lead to an increased rate of crystal growth provided that the degree of supersaturation is low. At high degrees of supersaturation, the concentration gradient near to the growing crystal is also high, so that an intensive flow of construction material to the crystal is available and the acoustic streaming has little effect on the crystal growth rate.

Studies on the kinetics of decomposition of supersaturated aluminium fluoride solutions in the presence of ultrasonic fields (25 kHz, 1 W cm^{-2}, and 800 kHz, 1 and 3 W cm^{-2}) showed that the ultrasound has no effect on the mechanism of crystallization. However, the rate of crystallization is increased by up to 7-fold (deposition at 90°C is reduced from 14 h to 2 h).[36] The effect is greatest at the lower frequency, which underlines the importance of cavitation in these processes.

Ultrasound has been reported to be of value in the hydrothermal synthesis of A-type zeolites by enabling cheap natural minerals such as kaolin to be used as the aluminosilicate source.[37]

At temperatures above the saturation points, crystals will dissolve; however, dissolution is not uniform across crystal faces and proceeds initially with the formation of etch pits. These etch pits are associated with defects within the lattice such as dislocations. Ultrasound at low

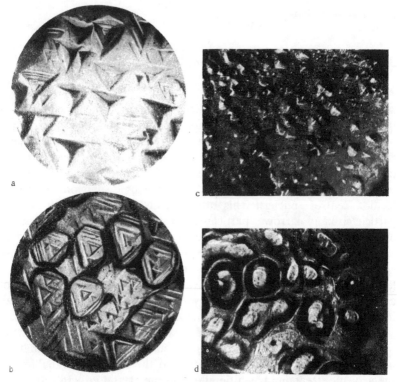

Fig. 2.6 Etch pits on potassium alum: (a) octahedral face, no ultrasound; (b) ultrasound; (c) rhombodecahedral face, no ultrasound; (d) ultrasound. (Reproduced with permission from A. P. Kapustin, *Effects of Ultrasound on the Kinetics of Crystallization*, Plenum Press, 1963.)

intensity has been shown to enhance the rate of development of these etch pits[38] (see Fig. 2.6). The rate of formation of etch pits is enhanced even further in plastically deformed crystals; for example Fig. 2.7 shows the increased density of etch pits formed along a scratch on a lithium fluoride plate.[2] Ultrasound at cavitation levels produces high pressure shockwaves, which on impinging on crystal faces can cause plastic deformation resulting in the production of microcracks which act as centres for the formation of etch pits.[2] At sufficiently high intensities, these shockwaves may cause crystal cleavage, resulting in increased surface area. The presence of surface defects such as dislocations and microcracks in addition to acting as sites for dissolution are also sites for a wide range of reactivity. This is particularly important in the field of heterogeneous catalysis. However, further discussion of this will be postponed until Section 2.5.

In addition to its role in crystallization processes, ultrasound has been shown to dramatically increase the rate of intercalation of a wide range of guest molecules into layered inorganic solids.[39,40] These rate enhancements were shown to be associated with a reduction in particle size of the host solid and with an increase in amount of surface damage. Thus, presonication of TaS_2 in toluene for only 15 min increased the rate of intercalation of n-hexylamine significantly and during this time the average particle size of TaS_2 was reduced from 60–90 μm to 5 μm. Further sonication had little effect on particle size and caused only a small further increase in rate of intercalation.

Ultrasound (17·5 kHz) has also been reported to produce significant increases in the rate of ion exchange of Ni^{2+} for Ca^{2+} in Y-type zeolites.[41]

2.3.2   Synthesis of inorganic complexes

High-intensity ultrasound has been shown to be beneficial in the reactions of metal carbonyl compounds. The results of sonication of $Fe(CO)_5$ are interesting because they are significantly different from the well-known photochemical and thermal reactions and also are able to go some way towards a better understanding of the mechanism of sonocatalysis. Thermolysis of $Fe(CO)_5$ gives finely divided iron powder, whereas ultraviolet photolysis gives mainly $Fe_2(CO)_9$ via the intermediate $Fe(CO)_4$. Sonolysis of $Fe(CO)_5$ in alkane solvents gives clusterification to $Fe_3(CO)_{12}$ and finely divided iron in a ratio which is

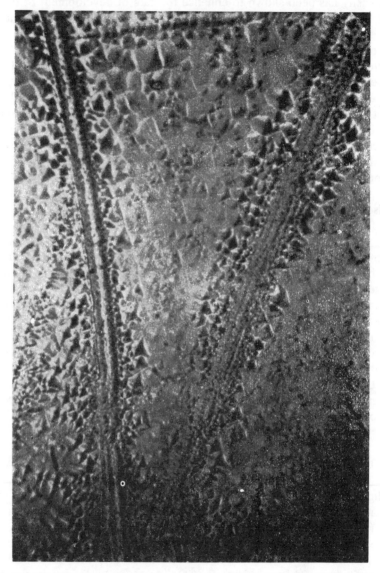

Fig. 2.7 Etch pits on a plastically deformed LiF crystal. (Reproduced with permission from A. P. Kapustin, *Effects of Ultrasound on the Kinetics of Crystallization*, Plenum Press, 1963.)

$$Fe(CO)_5 \rightarrow Fe(CO)_{5-n} + nCO$$

$$Fe(CO)_3 + Fe(CO)_5 \rightarrow Fe_2(CO)_8$$

$$2Fe(CO)_4 \rightarrow Fe_2(CO)_8$$ (Scheme 1)

$$Fe_2(CO)_8 + Fe(CO)_5 \rightarrow Fe_3(CO)_{12}$$

dependent on the solvent vapour pressure.[42,43] The clusterification to $Fe_3(CO)_{12}$ is favoured in solvents of high vapour pressure such as heptane, in which >82% is obtained, whereas in solvents of low vapour pressure such as decalin only 4·7% yield is obtained. These results are clearly related to the energetics of cavitation bubble collapse, which is inversely proportional to solvent vapour pressure. The clusterification, being the process of lower activation energy, is favoured by the lower-boiling solvents, which produce weaker cavitation. The formation of $Fe_3(CO)_{12}$ is considered to arise mainly from the multiple coordinatively unsaturated species $Fe(CO)_3$. The principal reactions are shown in Scheme 1. $Fe_2(CO)_9$ is not produced during the synthesis of $Fe_3(CO)_{12}$, and the sonolysis of $Fe_2(CO)_9$ yields only $Fe(CO)_5$ and finely divided iron. In the presence of added Lewis bases, such as phosphines and phosphites, substitution occurs to give $LFe(CO)_4$, $L_2Fe(CO)_3$, $L_3Fe(CO)_2$ and—only in solvents of low vapour pressure—small amounts of finely divided iron. The ratio of $LFe(CO)_4$ to $L_2Fe(CO)_3$ remains constant on prolonged sonolysis, which suggests that $LFe(CO)_4$ is inert to further sonochemical substitution. These observations are consistent with the same primary sonochemical event being responsible for the clusterification and substitution (Scheme 2).

Sonochemical substitution of other metal carbonyls, such as $Mn_2(CO)_{10}$, $Re_2(CO)_{10}$, $Cr(CO)_6$, $Mo(CO)_6$ and $W(CO)_6$, illustrates the generality of the technique.[44] In the case of the dimeric $Mn_2(CO)_{10}$ and $Re_2(CO)_{10}$ the substitution proceeds without the rupture of the metal–metal bond, which is more akin to thermal rather than photochemical substitution.

The metal carbonyl substitution reactions were used as probes to

$$Fe(CO)_5 \rightarrow Fe(CO)_{5-n} + nCO$$

$$Fe(CO)_4 + L \rightarrow LFe(CO)_4$$

$$Fe(CO)_3 + L \rightarrow LFe(CO)_3$$ (Scheme 2)

$$Fe(CO)_3L + L \rightarrow L_2Fe(CO)_3$$

$$(L = Ligand)$$

ascertain the nature of the hot spots produced during cavitation. Plots of the first-order rate constant against metal carbonyl vapour pressure are linear with a non-zero intercept, which suggests that there is a component of the reaction taking place in the gas phase (i.e. within the cavitation bubble) and the non-zero intercept suggests there is a pressure-independent reaction arising in the liquid phase, presumably in a thin spherical shell surrounding the bubble. Using sonochemical rate data and activation parameters from the literature, Suslick and co-workers,[45,46] calculated the effective temperature within the cavitation bubble to be 5200 ± 650 K and in the immediately surrounding liquid zone to be 1900 K, this liquid reaction zone was calculated to be 200 nm thick and to have a lifetime of less than 2 μs.

Ultrasound has also been shown to be beneficial in the synthesis of metal carbonyl anions.[47] Thus, reduction of transition-metal halides with sodium sand in the presence of CO at low pressure (1–5 atm) under prolonged sonication (probe 20 kHz, 100 W cm^{-2}) in THF gives reasonable yields of metal carbonyl anions: $[W_2(CO)_{10}]^{2-}$ 47%, $[Mo(CO)_{10}]^{2-}$ 54%, $[Nb(CO)_6]^-$ 51% and $[V(CO)_6]^-$ 35%. Metal halides with low solubility in THF give reduced yields. These results are quite remarkable, as such compounds are usually obtained by reduction of metal halides at high temperatures (100–300°C) and high CO pressures (100–300 atm) in an autoclave.

Another example of the formation of metal complexes by direct reaction with the ligand in an inert solvent is the reaction between copper and salicylanilide.[48] In this study, ultrasound (15–35 kHz, 20–30 W cm^{-2}) was shown to reduce the effective particle size of the copper powder to less than 5 μm and, interestingly, the reaction appeared to take place mainly on these small particles. It was suggested that the copper particles (<5 μm) were so small that they actually entered into the cavitation bubbles where, because of the high pressure and temperature, they would react instantaneously.

Ultrasound has been shown to facilitate the formation of alkali metal selenides and diselenides.[49] Thus, sonication (cleaning bath) of the reactions between the alkali metal with selenium in the presence of a small amount of naphthalene in THF gives up to 3-fold increase in rate compared with stirred control reactions.

2.4 Organometallic sonochemistry

Improvement in the synthesis of organometallic compounds has been a particularly fruitful area of sonochemical research.[50,51] In most cases the application of ultrasound has led to increases in rate and yield of products for existing methods, although in a few instances ultrasound has led to formation of new products. The most spectacular results have been observed in heterogeneous systems, particularly those containing solid phases where the effects of cavitation are known to be most pronounced.

The preparation of organometallic compounds of Li, Mg and Zn, which are of such great importance in organic synthesis, by direct reaction of the metal with alkyl or aryl halides traditionally requires scrupulously dry solvents and apparatus and the reactions are re-nowned for long induction periods. Various activation methods, such as the addition of iodine or 1,2-dibromoethane, are often necessary to initiate the reactions. However, very high yields of organolithium and Grignard reagents may be rapidly obtained in commercial undried solvents when the reactions are carried out in low-intensity acoustic fields generated by a simple ultrasonic laboratory cleaning bath.[52] Thus, *n*-propyl, *n*-butyl and phenyl lithium are obtained in >90% yield in a few minutes, while secondary and tertiary halides require slightly longer reaction times. However, improvements in the rate of these slower reactions may be obtained by increasing the acoustic intensity by means of an ultrasonic probe and by using a solvent which enables more powerful cavitation to occur, such as a THF/toluene mixture.[53] (The energetics of cavitation are inversely proportional to solvent vapour pressure.)

Similar reductions in induction times have been reported for the

Table 2.3 Preparation of butan-2-yl magnesium bromide in ether (50 kHz bath at 50°C)

Diethyl ether	Method	Induction time
Pure, dried (0·01% water)	non	6–7 mins
(0·01% ethanol)	u/s	<10 secs
Reagent grade (0·5% water)	non	2–3 hours (crushed)
(2·0% ethanol)	u/s	3–4 mins
50% Saturated (0·01% ethanol)	non	1–3 hours (crushed)
	u/s	6–8 mins

Source: Sprich, J. D. & Lewandos, G. S. (1982). *Inorg. Chem. Acta,* **76,** 1241.

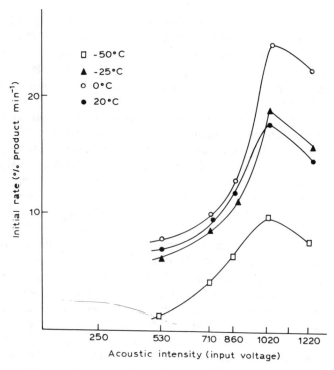

Fig. 2.8 Initial rate vs intensity of ultrasound for the Barbier reaction of benzaldehyde, *n*-heptyl bromide and lithium in THF.[55]

formation of Grignard reagents,[54] where the ultrasound appears to be effective in removing absorbed water from the magnesium surface, Table 2.3.

In a recent mechanistic study of the formation of organolithium compounds, under Barbier conditions, Luche *et al.*,[55] showed the rate of reaction is related to acoustic intensity. At acoustic intensities below the cavitation threshold no reaction is observed to occur, whereas at intensities greater than the cavitation threshold reaction rates increase with intensity up to a maximum value and thereafter the rate decreases with intensity (Fig. 2.8). This decrease in rate at high ultrasonic intensities was attributed to two factors, firstly, at high intensity cavitation of the liquid close to the radiation surface becomes so intense that the surface becomes shrouded in bubbles which act as an acoustic 'cushion' which diminishes the penetration of acoustic waves

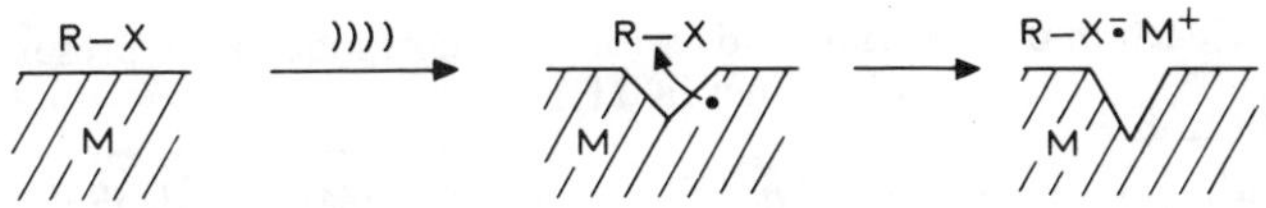

Scheme 3

into the liquid; secondly, at high intensities bubble growth may be so rapid that the bubble grows beyond the size range for transient cavitation before implosive collapse can occur.

The mechanism of formation of organolithium and organomagnesium reagents is generally considered to involve single electron transfer (SET),[56] from the metal to the substrate (Scheme 3). The sites on the metal surface where SET occurs most easily are at positions of surface defects such as dislocations, microcracks and positions of high stress. (See the earlier discussion on dissolution of metal salts.) Such defects are very easily produced by the hammering effects of shockwaves and micro jets of solvents produced by implosion of cavitation bubbles on the surface of the metal. Scanning electron microscopy of the surface of lithium after exposure to ultrasound in THF shows considerable surface roughening, and after exposure to *n*-heptylbromide in THF a large number of craters start to develop, rather similar to the etch pits shown in Fig. 2.6. Lowering the intensity of the ultrasound has the effect of reducing the number of these craters. As the initial rates of these reactions are inversely related to temperature, it may be inferred that the reactions are not mass-transport controlled, which suggests that the improvements in the rate which are obtained using ultrasound are related to the creation of the active sites for the SET to occur, i.e. the creation of lattice defects. Further confirmation of this was obtained by the observation that higher initial rates are obtained when the lithium is presonicated before the addition of the substrate provided the acoustic intensity is sufficiently low. Similar results are reported for reactions promoted by manganese.[57] From a practical point of view presonication of the metal is unnecessary provided a sufficiently high acoustic intensity is used.

Although ultrasound has been shown to depassivate and activate the surface of lithium under Barbier conditions, there are cases where the sonochemical Barbier reaction fails. For example, the reaction between *n*-octyl bromide with cyclohexanone in THF fails with lithium containing 0·02% sodium; however, 98% yields are obtained when lithium containing 2·0% sodium is used (Table 2.4). This was shown to

Table 2.4   Reaction of 2-octylbromide with cyclohexanone in the presence of lithium in THF

Metal	Catalyst (mol %)	Yield (% add. alcohol)
Li, 0·02% Na		Trace
Li, 2% Na		98
Li, 0·02% Na	CH_3ONa (5%)	98
	CH_3OLi (5%)	70
	$C_{16}H_{33}OLi$ (5%)	89
	Adogen (5%)	13

Source: Luche, J.-L., in Advances in Sonochemistry (1990). (Ref. 29.)

be due to the formation of sodium alkoxide in the early stages of the reaction.[29] From these studies it appears that the depassivation and creation of active sites on the metal surface are insufficient to promote reaction of alkyl halides with lithium in all cases.

The activation of metal surfaces by ultrasound has been found to be beneficial in the synthesis of organotransition metal complexes. The generation of the highly active Mg anthracene·3THF complex by reaction of Mg powder with anthracene in THF is facilitated by ultrasound (Scheme 4). The magnesium produced in this way is an excellent reducing agent for metal salts, and when the reduction is carried out in the presence of Lewis base ligands it is a useful route to organotransition metal complexes. For example η^5-cyclopentadienyl complexes, Cp_2M (M = V, Fe, Co), η^3-allyl complexes (M = Co, Ni), alkene complexes (M = Ni, Pd, Pt, Mo), and phosphine complexes (M = Pd, Pt).[58]

Ultrasonic activation of the reduction of Fe(dppe)Cl$_2$ by potassium in toluene/THF in the presence of pentamethylcyclopentadiene gives $Fe(C_5Me_5)(dppe)X$ (X = H, Cl).[59] This method represents a significant simplification in the methods of preparation of compounds of this type, as other methods require the generation of iron atoms.

Reduction of ruthenium chloride with zinc dust in methanol containing 1,5-cyclooctadiene in a 50 kHz ultrasonic bath gives a 93% yield of (η^6-1,3,5-cyclooctatriene)(η^4-1,5-cyclooctadiene)Ru compared with less than 35% yield using a two-stage non-ultrasonic method.[60]

Ultrasound also facilitates the production of arene radical anion salts of the alkali metals.[61–63] Thus, formation of sodium isobenzo-

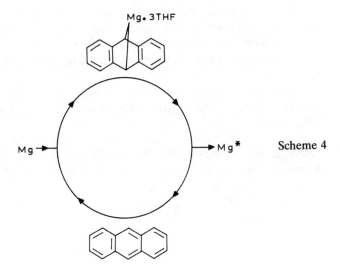

Scheme 4

quinoline is complete in 45 min using an ultrasonic method (probe 25 kHz) compared with 48 h using conventional mechanical stirring.

Improvements in yields and reaction rates have been observed in the synthesis of organoaluminium compounds.[64,65] Thus, an 82% yield of ethyl aluminium sesquiiodide was obtained in 2 h at 40°C using an ultrasonic cleaning bath (43 kHz, 180 W) compared with only a 21% yield in a stirred control reaction.

Insonation (25 kHz probe) has led to significant improvements in the synthesis of organoboranes via the generation *in situ* of Grignard reagents.[66] This method is particularly useful for those systems which are traditionally slow, such as those with hindered boranes. For example, (1-naphthyl)$_3$B is obtained in 93% yield in 25 min compared with a 91% yield in 24 h in the normal non-ultrasonic method.

There have been many reports of the use of ultrasound in the direct formation of organozinc reagents.[67–70] However, a more reliable method involves trans-metallation of zinc halides with in-situ ultrasonically generated organolithium compounds (eqn 2.2).[71]

$$RX + Li \xrightarrow{\;)))\;} RLi \xrightarrow{\;ZnBr_2\;} R_2Zn \qquad (2.2)$$

While good yields of diarylzinc can be obtained by a one-pot procedure using aryl halides, zinc bromide and lithium wire in THF in an ultrasonic bath (50 kHz), more reliable results are obtained for dialkylzinc when a more intense acoustic field provided by a sonic horn is used.

Organocopper reagents can also be prepared by trans-metallation of copper(I) halides with in-situ ultrasonically generated organolithium reagents.[72]

There have been many reports of the use of ultrasound in the synthesis of organosilanes, organogermanes and organostannanes. Significant improvements in the synthesis of hexaalkyldisilanes, germanes and stannanes are obtained in the Wurtz-type homocoupling of chlorotrialkylsilanes using lithium in THF in an ultrasonic cleaning bath (eqn 2.3).[73]

$$R_3MCl \xrightarrow[)))]{Li,THF} R_3MMR_3 \qquad (2.3)$$

M = Si, Ge, Sn
R = Me, Et, Bu, Ph

Reaction of dichlorodialkylsilanes under these conditions leads mainly to cyclic polysilanes (eqn 2.4)

$$R_2SiCl_2 \xrightarrow[)))]{Li,THF} (R_2Si)_n \qquad (2.4)$$

R = Me, $n = 6$
R = Ph, $n = 4$

However, the reaction of dichlorodialkylsilanes with sodium in non-polar solvents (toluene) under insonation (25 kHz probe) gives good yields of high-molecular-weight polysilanes at ambient temperature.[74]

With highly hindered dichlorodiorganosilanes it is possible to generate silene intermediates using lithium under insonation in THF. These silenes readily insert into Si—H bonds,[75] (eqn 2.5)

$$(t\text{-}C_4H_9)_2SiCl_2 \xrightarrow[)))]{Li,THF} [(t\text{-}C_4H_9)_2Si:] \xrightarrow{R_3SiH} R_3Si(t\text{-}C_4H_9)H \quad (2.5)$$

This method has also been used to prepare the mesitylene stabilized silene, tetramesityldisilene[76] (eqn 2.6).

$$(Mes)_2SiCl_2 \xrightarrow[\text{)))}]{\text{Li,THF}} \begin{array}{c} Mes \\ Mes \end{array} Si{=}Si \begin{array}{c} Mes \\ Mes \end{array} \qquad (2.6)$$

$$Mes = \text{[structure]}$$

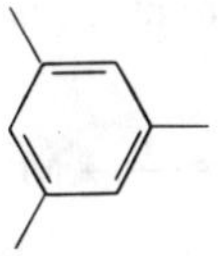

Cross coupling of allyl halides with chlorotrialkylstannanes in the presence of magnesium is a convenient route to allytrialkylstannanes (eqn 2.7).[77]

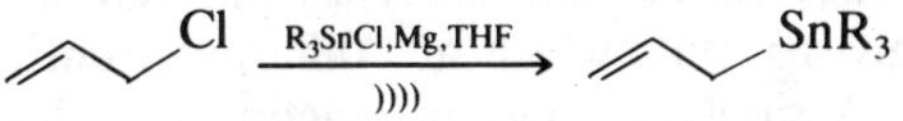

$$\text{\textbackslash\_Cl} \xrightarrow[\text{))))}]{R_3SnCl,Mg,THF} \text{\textbackslash\_SnR_3} \qquad (2.7)$$

These reactions are regiospecific and stereospecific and are complete in less than 1 h, whereas in the absence of ultrasound substantial amounts of homocoupled products are obtained.

The addition of sonochemically in-situ generated organosilylzinc to unsaturated systems has been widely exploited in organosilicon chemistry (eqn 2.8).[67]

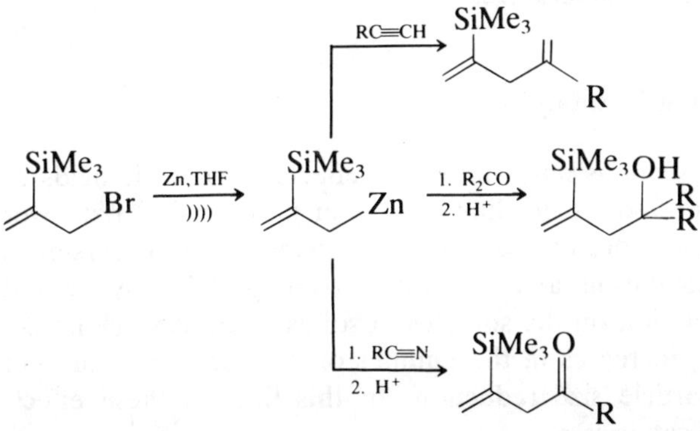

$$(2.8)$$

Ultrasound facilitates the reduction of R_3MX (X = halogen, alkoxy, amino; R = alkyl, aryl; M = Si, Ge, Sn) with lithium hydride to give the corresponding triorganosilane[78] (eqn 2.9).

$$R_3MX \xrightarrow[\text{)))}]{LiAlH_4} R_3MH \qquad (2.9)$$

Ultrasound has also been shown to give significant improvements in the hydrosilation of alkenes using a Pd/C catalyst.[79]

A novel route to organoselenium and organotellurium compounds involves ultrasonic acceleration of the electroreduction of Se and Te to their dianions followed by reaction with electrophiles RX[80] (eqns 2.10, 2.11)

$$M + 2\varepsilon \xrightarrow{\;))) \;} M^{2-} \xrightarrow{\;RX\;} MR_2 \qquad (2.10)$$

$$2M + 2\varepsilon \xrightarrow{\;))) \;} M_2^{2-} \xrightarrow{\;RX\;} M_2R_2 \qquad (2.11)$$

$$M = Se, Te$$

We have already noted that sonolysis of solutions of $Fe(CO)_5$ in hydrocarbons gives results which are different from either photolysis or thermolysis. A similar change in reaction path has been reported for the heterogeneous reaction of $Fe_2(CO)_9$ and anthracene. The product of the sonochemical reaction is $Fe_2(CO)_6(C_{14}H_{10})$, whereas the thermal reaction gives $Fe(CO)_3(C_{14}H_{10})$.[81]

Significant improvements in the synthesis of the synthetically important $(\eta^3$-allyl)$Fe(CO)_3$[82] and $(\eta^4$-diene)$Fe(CO)_3$ complexes[83] are also obtained by sonolysis of slurries of $Fe_2(CO)_9$ with the ligand in benzene at room temperatures.

2.5 Ultrasound in catalysis

The previous sections have noted the effects of acoustic fields on the crystallization of solids, leading to smaller particle size, narrower size distribution and more equiaxed crystals owing to the dispersing action of acoustic cavitation and acoustic streaming. We have noted the effects of cavitation on the surfaces of solids, such as work hardening, depassivation, increases in the number of initiation sites for chemical action and particle size reduction. In this Section these effects are considered in catalysis.

2.5.1 Heterogeneous catalysis

2.5.1.1 Catalyst preparation   Reduction of metal salts in the presence of acoustic fields has led to catalysts with improved activity over those produced with conventional mechanical agitation.

Thus, insonation during the preparation of platinum blacks by reduction of aqueous solutions of platinum metal salts with formaldehyde gives blacks with up to 3-fold increase in activity over those produced with stirring at 1000 r.p.m. in the hydrogenation of alkenes, the decomposition of hydrogen peroxide and the oxidation of ethanol.[84] Platinum blacks produced in the presence of ultrasound showed a 62% increase in surface area and a 98% increase in magnetic susceptibility, which was attributed to an increase in the amount of atomic phase, which is more catalytically active than the crystalline metal. An interesting but unexplained frequency effect was also observed during this study. The most active platinum blacks were obtained at high frequency (3 MHz), whereas with palladium the highest activity was obtained at low frequency (20 kHz)—see Table 2.5. However, the acoustic intensity in these studies was quite low. Insonation of preformed platinum blacks, however, leads to a 38% reduction in surface area and up to a 50% increase in crystal size.[85] From these results it is clear that the ultrasound is most effective during the crystallization process. This is probably due to its ability to increase the rate of nucleation by dispersion of nuclei.

Insonation (25 kHz, $0.3 \, \text{W cm}^{-2}$) during the precipitation of cobalt and nickel oxalates also leads to a 50% reduction in particle size, lower porosity, increased specific surface area and increased bulk density.[86,87] Subsequent reduction of the oxalate to the metal gives a nickel catalyst with an 87% increase in activity in the hydrogenation of benzene to cyclohexane. The corresponding cobalt catalyst showed a 14% increase in activity in this reaction. Similarly, nickel and copper catalysts prepared by precipitation from aqueous solutions of the nitrates in an acoustic field (550 kHz) followed by reduction to the metal show up to 40% increase in activity in the hydrogenation of oils.[88]

Table 2.5 Increase in activity of platinum and palladium blacks obtained in an ultrasonic field (activity of the black obtained non-ultrasonically = 1·0)

Catalytic process	Pt black			Pd black		
	2 MHz	*548 kHz*	*20 kHz*	*3 MHz*	*548 kHz*	*20 kHz*
Decomposition of H_2O_2	2·60	1·75	1·35	1·00	2·00	3·20
Hydrogenation of hex-1-ene	1·38	1·04	1·13	1·75	1·40	1·30
Oxidation of ethanol	1·40	0·84	0·74	0·79	1·35	2·40

Source: Maltsev, A. N. (1976). *Russ. J. Phys. Chem.*, **50**, 995.

There have been several reports of the use of ultrasound in the preparation of metal oxide catalysts. For example, a mixed chromium–molybdenum oxide catalyst precipitated in an ultrasonic field followed by calcination shows enhanced activity in the oxidation of methanol to formaldehyde.[89,90] This catalyst had a 15–20% increased dispersity compared with a control sample. Application of ultrasound after the precipitation stage leads to agglomeration, surface area reduction and a reduction in activity, presumably owing to the increase in the number of molecular collisions.

Silica gel precipitated in the presence of ultrasonic fields (90 kHz) shows increased activity in the decomposition of hydrogen peroxide despite reductions of 18% in surface area and 30% in pore volume. Electron microscopy of the insonated gel revealed a large increase in the number of surface cracks, which are claimed to be the active centres for the reaction.[91] During this study it was noted that both the activation energy and the frequency factor for the reaction on the insonated gel were increased.

Similar effects were noted during the decomposition of hydrogen peroxide over a series of alumina-supported metal-oxide catalysts (Table 2.6). This suggests that insonation has produced both an increase in the number and a change in type of active sites.[92]

Increases in the surface concentration of Cr(VI) species and in specific surface area are claimed to be the principal reasons for a 30%

Table 2.6 Alumina-supported catalysts for H_2O_2 decomposition

Catalyst	$E\dagger$ $(kcal\,mol^{-1})$	$Log\,A$	Surface area (m^2g^{-1})	Rate constant $50°$ $(h^{-1}g^{-1})$
Al_2O_3 support	—	—	220	—
Cr_2O_3/Al_2O_3				
Ultrasound	14·8	10·73	135	4·17
Normal	13·3	9·50	108	3·37
MnO_2/Al_2O_3				
Ultrasound	7·0	5·85	120	13·0
Normal	5·64	4·84	107	10·6
Co_2O_3/Al_2O_3				
Ultrasound	7·9	5·60	125	1·8
Normal	7·0	4·94	105	1·6

Source: Ranganathan *et al.*[92]

increase in the rate of decomposition of hydrogen peroxide activity of Cr_2O_3 gels precipitated in the presence of acoustic fields.[93]

Anodic oxidation of $MnSO_4$ in the presence of ultrasound gives MnO_2 which also shows increased activity in the decomposition of hydrogen peroxide.[94] This increase in activity was also attributed to increases in specific surface area and increases in the number of surface defects.

Ultrasound is also reported to be beneficial in the dispersion of catalysts on to inert supports such as silica, alumina and layered inorganic solids. Thus, reduction of aqueous ammonium hexachloroplatinate containing suspended silica gel in an ultrasonic field (440 kHz, 5 W cm^{-2}) gives an 80% increase in surface area of metal over a mechanically stirred control reaction.[95]

2.5.1.2 Catalyst activation The ability of ultrasound to clean and activate surfaces has been exploited successfully in catalysis. Insonation (20 kHz, 50 W cm^{-2}) of ordinary 5 μm nickel powder in octane increases its activity as an alkene hydrogenation catalyst by a factor of 10^5.[96] This represents a significant simplification in procedure compared with the usual methods for the preparation of active nickel powders. The initial rate of hydrogenation rises to a maximum with nickel which has been presonicated for 60 min and thereafter decreases as the extent of presonication is increased (Fig. 2.9). Scanning electron microscopy shows that there is no change in particle size during sonication; however, there is a marked change in surface morphology. Initially the surface is crystalline but becomes smooth after sonication, there is also an increase in agglomeration on prolonged sonication. This is no doubt the principal reason for the decrease in rate of hydrogenation after prolonged sonication (Fig. 2.9). Both of these effects are attributed to a large increase in number of interparticle collisions in the acoustic field. The most striking effect of sonication on the activation process was revealed by Auger electron spectroscopy. The initial nickel powder is covered in a thick oxide coating (surface Ni:O ratio = 1·0) extending to a depth of 25 nm (250Å), however, after sonication for 1 h most of the oxide has been removed (surface Ni:O ratio = 2·0) and the depth is reduced to 5 nm.

Insonation of Raney nickel has been shown to increase the activity in catalysis of H/D exchange in carbohydrates and glycosphingolipids.[97] An extensive study of unsonicated and sonicated Raney nickel by X-ray photoelectron spectroscopy, Auger electron

54   *J. Lindley*

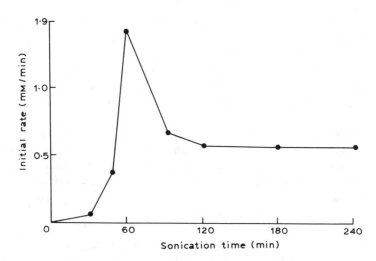

Fig. 2.9 Activation of nickel powder as an alkene hydrogenation catalyst using ultrasound at 20 kHz, 50 W cm^{-2}.[96]

spectroscopy, argon ion sputtering, static secondary-ion mass spectroscopy and surface area measurements reveals that the acoustic field increases and develops the catalytic sites, removes passivating impurities and causes an elemental redistribution within the bulk catalyst.[98]

Insonation (20–500 kHz; 4–20 W cm^{-2}) of a titanium trichloride Ziegler–Natta catalyst suspended in octane leads to a reduction in particle size from 15–40 μm to 0·1–5 μm.[99] This catalyst gives rise to crystalline polymers with a more uniform molecular weight distribution than those obtained without ultrasonic treatment. Similar results are reported for other Ziegler–Natta catalysts.[100,101]

The ability of ultrasound to clean surfaces has been exploited in the area of catalyst regeneration. Thus, a brass catalyst used in the production of acetone from isopropanol is regenerated to 83% of its original activity by insonation (18–22 kHz, 6–20 W cm^{-2}) in a bath containing a solution of sulphuric acid, nitric acid and sodium dichromate;[102] non-ultrasonic treatment leads to only 63% regeneration.

Insonation is reported to restore completely the specific surface area, porosity and activity of a palladium–alumina catalyst (APK-Z) for the removal of nitrogen oxides from waste gases.[103,104]

A nickel-molybdenum hydrocarbon-cracking catalyst is reported to be regenerated by oxidation in air followed by insonation in low-viscosity oil.[105] Similarly, a TiO_2–V_2O_5 catalyst for denitrification of flue gases is reactivated by insonation in water for 0·5 h.[106]

2.5.1.3 Ultrasound during catalysis Insonation of catalytic systems has been widely reported; benefits claimed include improved mass transport, generation and renewal of active sites, and desorption of products and poisons from the catalyst surface.

Gas–solid systems Clearly, in the absence of a liquid phase the major benefits of ultrasound can only be due to improved thermal and mass transport. Lintner and Hanesian[107] reported that insonation (26 kHz and 39 kHz, 0·05–1·3 W cm^{-2}) during the conversion of cumene to propene and benzene in the temperature range 617–839 K over a silica–alumina cracking catalyst leads to a 40% increase in mass transfer coefficient when external bulk diffusion controls the reaction and up to 160% increase in rate constant when surface reaction and internal pore diffusion control the reaction.

Improved mass transport was found to be the major factor for a 50% increase in the rate of thermal decomposition of formic acid at a nickel filament which was observed in a standing-wave field (13·5 kHz, 0·05 W cm^{-2}).[108] However, under these conditions only a 14% increase was observed in the rate of thermal decomposition of ammonia and the hydrogenation of ethene was found to be unaffected.

Liquid–liquid–gas systems Significant increases in turnover numbers are observed in the hydroformylation of alkenes using a water-soluble rhodium catalyst in the presence of ultrasonic fields (35–40 kHz).[109] Thus, using a 1:1 ratio of CO to H_2 at 2·5 MPa, 1-hexene and an aqueous solution containing $HRhCl(CO)[P(C_6H_4SO_3H)_3]_3$ aldehydes were produced with a turnover number of 11·34 in the ultrasonic field compared with only 3·24 with stirring at 500 r.p.m.

Liquid–solid–gas systems There are many reports of the use of ultrasound during hydrogenations over a variety of catalysts. For example, hydrogenation of acrylic acid in the presence of platinum metal blacks at 25°C proceeds with 3-fold increase in rate in ultrasonic fields (440 kHz, 5 W cm^{-2}) compared with stirring at 1000 r.p.m.[110]

J. Lindley

Table 2.7 Ammonia synthesis

Catalyst	Yield of NH_3 $(g\ cm^{-3})$	Increase with catalyst
None	1·2	—
Pt black (prepared in situ)	6·0	5
Pt black	3·0	2·5
Rh black (prepared in situ)	9·0	7·5
Rh black	4·3	3·6
Pd black (prepared in situ)	4·0	3·3
Pd black	2·4	2·0

Source: Maltsev, A. N. (1976). *Russ. J. Phys. Chem.*, **50**, 995.

Hundredfold increases in the rate of hydrogenation of soya-bean oil with copper chromite or Nysel catalysts are reported using ultrasound (20 kHz) in a flow reactor.[111,112] Ultrasonically enhanced hydrogenation using in-situ generated hydrogen from compounds such as formic acid (Pd/C catalyst),[113] hydrazine (Pt/C catalyst),[114] and water (Zn/Ni catalyst)[115] also fall into this category.

Although the gas phase is inert in these reactions, it plays a vital role in the cavitation process which activates the catalyst and leads to the generation of radicals which are believed to be important in these hydrogenations.

$$H_2O \rightarrow H + OH$$
$$OH + H_2 \rightarrow H + H_2O$$

NO CATALYST:

$$N_2^* + H \rightarrow NH + N$$
$$N + H_2 \rightarrow NH + H \text{ etc.}$$
$$NH + H_2 \rightarrow NH_3 \qquad\qquad \text{(Scheme 5)}$$

CATALYST:

$$N_2^* + H_{2(ads)} \rightarrow N_{(ads)} + NH_{2(ads)}$$
$$N_2^* + H_{2(ads)} \rightarrow N_{(ads)} + NH_{(ads)} + H_{(ads)}$$
$$NH_{2(ads)} + H_2 \rightarrow NH_{3(ads)} + H_{(ads)}$$
$$NH_{(ads)} + H_2 \rightarrow NH_{3(ads)}$$

Substantial improvements in the Pt/C catalysed hydrosilation of alkenes are reported for reactions carried out in an ultrasonic cleaning bath at 30°C, this is the lowest temperature ever reported for Pd–C catalysed hydrosilations.[116,117]

Ultrasound (500 kHz, 4–5 W cm^{-2}) produces an acceleration of the rate of formation of ammonia from nitrogen and hydrogen in aqueous solution (Table 2.7)[110] In the absence of catalyst, ammonia production is thought to be initiated by the $^3\Sigma_u$ excited state of nitrogen formed within the cavitation bubbles and by hydrogen radicals formed by homolysis of water (Scheme 5).

Insonation of solid barium hydroxide leads to substantial improvements in yield, rates and catalyst concentration for a variety of reactions of aromatic aldehydes, such as the Cannizzaro, Michael addition, Claisen–Schmidt, aldol condensation and the Wittig–Horner reactions.[118–120]

Some results for the Cannizzaro disproportionation of 4-chlorobenzaldehyde are shown in Table 2.8. It is noteworthy that no reaction occurs at 25°C in the absence of catalyst and in the absence of ultrasound. By means of selective site poisons it was established that

Table 2.8   Barium hydroxide-catalysed reactions. Cannizzaro reaction of 4-chlorobenzaldehyde. Sonication for 10 min at 25°C

Catalyst	Mass (g)	Yield %
Activated Ba(OH)$_2$	0·0	0
Activated Ba(OH)$_2$	0·005	100
Activated Ba(OH)$_2$	0·010	100
K$_2$CO$_3$	0·010	100

Catalyst	Reduction centres (equiv. g^{-1})	Moles dinitrobenzene per g catalyst	Yield (%)
Activated Ba(OH)$_2$	3·5 × 10^{-5}	—	100
Activated Ba(OH)$_2$	3·5 × 10^{-5}	5 × 10^{-5}	0

Catalyst	Base centres	Moles TBMPHE per g catalyst	Yield %
Activated Ba(OH)$_2$	6·5 × 10^{-6}	—	100
Activated Ba(OH)$_2$	6·5 × 10^{-6}	6·5 × 10^{-6}	100

J. Lindley

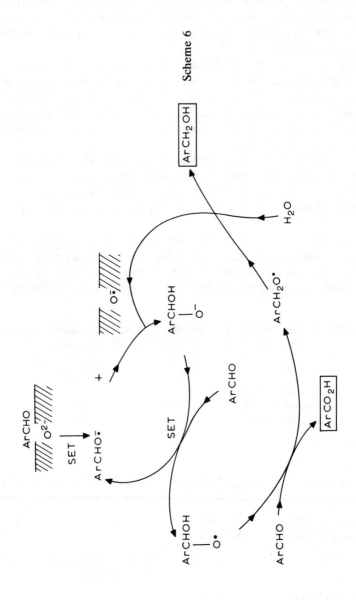

Scheme 6

$$M(CO)_n \rightarrow M(CO)_m + (n - m)CO$$

$$M(CO)_m + \text{1-alkene} \rightarrow M(CO)_x(\text{1-alkene}) + (m - x)CO$$

$$M(CO)_x(\text{1-alkene}) \rightarrow M(CO)_x(H)(\pi\text{-allyl}) \qquad \text{(Scheme 7)}$$

$$M(CO)_x(\pi\text{-allyl}) \rightarrow M(CO)_x(\text{2-alkene})$$

$$M(CO)_x(\text{2-alkene}) + \text{1-alkene} \rightarrow M(CO)_x(\text{1-alkene}) + \text{2-alkene}$$

the reaction occurs at reducing sites on the catalyst and a mechanism involving single electron transfer from the catalyst to the substrate initiates the catalytic cycle (Scheme 6).

2.5.2 Homogeneous catalysis

We noted earlier that sonication at cavitational intensities can lead to ligand dissociation in metal carbonyl complexes. The resulting coordinatively unsaturated species are excellent catalysts for the isomerization of 1-alkenes to internal isomers. Initial turnover numbers are as high as 100 and represent rate enhancements of 10 times over thermal controls.[42,43] The sonocatalytic and photocatalytic activities of these carbonyls are in general accord. While the exact nature of the catalytic species is unknown, a mechanism involving π-allyl intermediates is proposed (Scheme 7).

Another example of homogeneous sonocatalysis involving metal carbonyls is the sonochemical oxidation of alkenes by oxygen in the presence of $Mo(CO)_6$ to give 1-enols and epoxides.[121] However, in this reaction the primary sonochemical event is considered to be the homolysis of the allylic C—H bond caused by cavitation (Scheme 8).

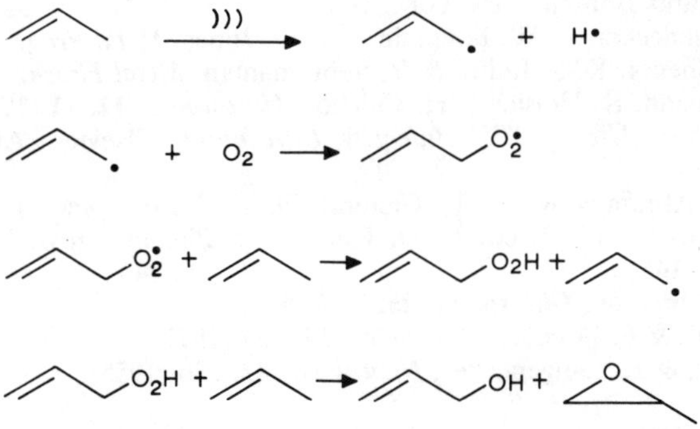

Scheme 8

2.6 Conclusion

Although sonochemistry is over sixty years old, the recent growth of interest has undoubtedly been aroused by the spectacular improvements obtained in synthesis, especially in those reactions involving heterogeneous systems.

Sonochemical depassivation and activation of metals has been particularly successful in enabling reactions to occur under less severe conditions and at lower temperatures. The intense mixing due to cavitation bubble collapse and acoustic streaming is of particular value for those reactions which are mass-transport controlled. The ability of ultrasound to effect crystal nucleation and growth is not new, especially in the field of metal technology, and there appears to be scope for its application to inorganic systems. The ability of ultrasound to reduce diffusion layers is of particular significance in electrochemistry. The application of ultrasound in inorganic chemistry is an area which has seen little exploitation, apart from metal carbonyl chemistry, and has further potential in synthesis and homogeneous catalysis. The application of ultrasound in the preparation and activation of heterogeneous catalysts appears to be benefical in many cases, although the continuous application of ultrasound to some catalytic systems can lead to catalyst deactivation by causing agglomeration and reduction in pore volumes.

2.7 References

1. P. J. Lorimer & T. J. Mason, Reviews on physical principles of ultrasound. *Chem. Soc. Rev.*, **16**, 239 (1987).
2. A. P. Kupustin, *Effects of Ultrasound on the Kinetics of Crystallization.* Consultants Bureau, New York, 1963.
3. B. Langenecker, C. W. Fountain & V. O. Jones, *Metal Prog.*, 97 (1964).
4. B. A. Shenoi, K. S. Indira & R. Subramanian, *Metal Finish.*, 41 (1970).
5. R. Pohlman, K. Heisler & H. Chichos, *Ultrasonics,* **12**, 11 (1974).
6. A. Puskar, *Use of High Intensity Ultrasound,* Elsevier, Amsterdam, 1982.
7. O. V. Abramov & V. I. Teumin, *Physical Principles of Ultrasonic Technology,* Vol. 2, ed. L. D. Rozenberg, Plenum Press, New York, 1973, p. 145.
8. O. V. Abramov, *Ultrasonics,* **25**, 73 (1987).
9. J. Seifect & F. Fischer, *Ultrasonics,* **15**, 154 (1977).
10. F. Blaha & B. Langenecker, *Naturwiss.,* **42**, 536 (1955).

11. W. P. Mason, *Physical Acoustics and the Properties of Solids.* University of Princeton Press, Cambridge, MA, 1958.
12. R. E. Green, *Ultrasonics*, **13**, 117 (1975).
13. E. Neppiras, *J. Appl. Phys.*, **11**, 143 (1960).
14. R. A. Geckle, *Metal Finishing Guide Book*, 144 (1980).
15. A. Peslo, *Ultrasonics*, **22**, 37 (1984).
16. R. Pohlman, K. Heisler & M. Chicos, *Ultrasonics*, 11 (1974).
17. P. Boudjouk, D. P. Thompson, W. H. Ohrbom & B. H. Han, *Organometallics*, **5**, 1257 (1986).
18. R. D. Rieke, *Acc. Chem. Res.*, **10**, 301, (1977).
19. H. Bonnemann, B. Bogdanovic, D. W. He & B. Spliethoff, *Angew. Chem. Int. Ed*, **22**, 728 (1983).
20. J.-L. Luche, C. Petrier & C. Dupuy, *Tetrahedron Lett.* **25**, 753 (1984).
21. A. J. Fry & D. Herr, *Tetrahedron Lett.* **19**, 1721 (1978).
22. J. Lindley, P. J. Lorimer & T. Mason, *Ultrasonics*, **24**, 292 (1986).
23. B. A. Shenoi, K. S. Indira & R. Subramanian, *Metal Finish.*, 57 (1970).
24. S. M. Kochergin & G. Y. Vyaseleva, *Electrodeposition of Metals in Ultrasonic Fields*, Consultants Bureau, New York, 1966.
25. S. I. Pugachev & N. G. Semenova, *Vozdectsev Mostrichn Ul'trazvoka Mezh. Pov. Met.*, ed A. I. Manokhin, (1986) p. 95.
26. R. Walker & J. F. Clements, *Metal Finish. J.*, 101 (1970).
27. W. Wolfe, H. Chessin, E. Yeager & F. Hovorka, *J. Electrochem. Soc.*, **101**, 590 (1954).
28. M. P. Drake, *Trans. Inst. Metal Finish.*, **58**, 67 (1980).
29. J.-L. Luche, *Advances in Sonochemistry*, ed. T. J. Mason, JAI Press, 1990.
30. B. A. Shenoi, K. S. Indira & R. Subramanian, *Metal Finish.*, 41 (1970).
31. C. T. Walker & R. Walker, *Nature*, **244**, 141 (1973).
32. C. T. Walker & R. Walker, *Ultrasonics*, 79 (1975).
33. E. B. Kenahen & D. Schlain, *Proc. Am. Electropl. Soc.*, **47**, 158 (1960).
34. R. Walker & S. A. Halagan, *Plating and Surface Finish*, **72**, 69 (1985).
35. W. T. Richards & A. L. Loomis, *J. Am. Chem. Soc.*, **49**, 3086 (1927).
36. Y. T. Tyurin & S. I. Rempel', in *Crystallization Processes*, ed. N. N. Sirota & F. K. Gorski. Consultants Bureau, New York, 1966, p. 155.
37. Y. Ueda, H. Sekiguclin & H. Shirahata, Jap Patent 60-226413, 1985.
38. A. P. Kapustin, in *Crystallization Processes*, ed. N. N. Sirota & F. K. Gorski, Consultants Bureau, New York, 1966, p. 135.
39. K. S. Suslick, D. J. Casadonte, M. L. H. Green & M. E. Thompson, *Ultrasonics*, **25**, 26 (1987).
40. K. S. Suslick, D. J. Casadonte, M. L. H. Green & M. E. Thompson, *J. Chem. Soc. Chem. Commun.*, 901 (1987).
41. Y. A. Egiazarov, E. Y. Ustilovskaya, A. K. Kudoyavstev, Y. A. Paushkin & E. G. Sovochkin, *Ves. Akad. Nauk. B USSR Ser. Khim. Nauk*, 10 (1975).
42. K. S. Suslick, P. F. Schubert & J. W. Goodale, *J. Am. Chem. Soc.*, **105**, 7324 (1983).
43. K. S. Suslick, J. W. Goodale, P. F. Schubert & H. H. Wang, *J. Am. Chem. Soc.*, **105**, 5781, (1983).

44. K. S. Suslick & P. F. Schubert, *J. Am. Chem. Soc.*, **105**, 6042 (1983).
45. K. S. Suslick, R. E. Chine, Jr, D. A. Hammerton, *J. Am. Chem. Soc.*, **108**, 5641 (1986).
46. K. S. Suslick & D. A. Hammerton, *Proc. Ultrasonics Int. Conf.*, London. Butterworths, 1985, p. 231.
47. K. S. Suslick & R. E. Johnson, *J. Am. Chem. Soc.*, **106**, 6856 (1984).
48. A. S. Kuzharov, L. A. Vlasenko & V. V. Suchkov, *Russ. J. Phys. Chem.*, **58**, 542 (1984).
49. D. P. Thompson & P. Boudjouk, *J. Org. Chem.*, **53**, 2109 (1988).
50. K. S. Suslick, *Adv. Organometall. Chem.*, **25**, 73 (1986).
51. J. Lindley & T. J. Mason, *Chem. Soc. Rev.*, **16**, 275 (1987).
52. J.-L. Luche & J. C. Damanio, *J. Am. Chem. Soc.*, **102**, 7926 (1980).
53. C. Petrier, J. C. de S. Barabara, C. Dupuy & J.-L. Luche, *J. Org. Chem.*, **50**, 5761 (1985).
54. J. D. Sprich & G. S. Lewandos, *Inorg. Chem. Acta*, **76**, 1241 (1982).
55. J. C. de Barbosa, C. Petrier, J.-L. Luche, *J. Org. Chem.*, **51**, 55 (1988).
56. H. M. Walborsky & M. S. Arnoff, *J. Organometall. Chem.*, **51**, 55 (1973).
57. B. Pugin & A. Turner, *Advances in Sonochemistry*, JAI Press, 1990.
58. H. Bonnermann, B. Bogdanovic, R. Brinkman, W. He & B. Spliehoff, *Angew. Chem. Int. Ed.*, **22**, 728 (1983).
59. C. Roger, P. Marseille, C. Salns, J.-R. Hamon & C. Lapinte, *J. Organometall. Chem.*, **336**, C13 (1987).
60. K. Itoh, H. Nagashima, T. Ohshima, N. Ohshima & N. Nishiyama, *J. Organometall. Chem.*, **272**, 179 (1984).
61. W. Slough & A. R. Ubbelhode, *J. Chem. Soc.*, 918 (1957).
62. M. W. T. Pratt & R. Helsby, *Nature*, **184**, 1694 (1959).
63. T. Azuma, S. Yanagida, H. Sakurai, S. Sasa & K. Yoshino, *Synth. Commun.*, **12**, 137 (1982).
64. A. V. Kurchin, R. A. Nurushev & G. A. Tolstikov, *Z. Obschch. Khim.*, **53**, 2519 (1983).
65. K. F. Lion, P. H. Yang & Y. T. Lin, *J. Organomet. Chem.*, **294**, 145 (1985).
66. H. C. Brown & U. S. Racherla, *Tetrahedron Lett.*, **26**, 4311 (1985).
67. P. Knockel & J. F. Normant, *Tetrahedron Lett.*, **25**, 1475 (1984).
68. B. H. Han & P. Boudjouk, *J. Org. Chem.*, **47**, 751 (1982).
69. T. Kitazume & N. Ishikawa, *J. Am. Chem. Soc.*, **107**, 5186 (1985).
70. T. Kitazume & N. Ishikawa, *Chem. Lett.*, 137 (1982).
71. J.-L. Luche, C. Petrier, J.-P. Lansard & A. E. Greene, *J. Org. Chem.*, **48**, 3837 (1983).
72. J.-L. Luche, C. Petrier, A. L. Geinal & N. Zirka, *J. Org. Chem.*, **47**, 3805 (1982).
73. P. Boudjouk & B. H. Han, *Tetrahedron Lett.*, **22**, 3813 (1981).
74. H. K. Khim & K. Matyjaszawski, *J. Am. Chem. Soc.*, **110**, 3321 (1988).
75. P. Boudjouk, *J. Chem. Educ.*, **63**, 427 (1986).
76. P. Boudjouk, B. H. Han & K. R. Anderson, *J. Am. Chem. Soc.*, **104**, 4992 (1982).
77. Y. Naruta, Y. Nishigaichi & K. Marvyama, *Chem. Lett.*, 1857 (1986).

78. E. Lubevics, V. N. Grevorgyan & Y. S. Goldberg, *Tetrahedron Lett.*, **25**, 4383 (1984).
79. B. H. Han & P. Boudjouk, *Organometallics*, **2**, 769 (1983).
80. B. Gautheron, G. Tainturer and C. Degrand, *J. Am. Chem. Soc.*, **107**, 5579 (1985).
81. M. J. Begley, S. G. Puntambekar & A. H. Wright, *J. Chem. Soc. Chem. Commun.*, 1251 (1987).
82. A. M. Horton, D. M. Hollinshead & S. V. Ley, *Tetrahedron*, **40**, 1737 (1984).
83. S. V. Ley, C. M. R. Low & A. D. White, *J. Organometall. Chem.*, **302**, C13 (1986).
84. L. Wen-chou, A. N. Mal'tsev & N. I. Kobosev, *Russ. J. Phys. Chem.*, **38**, 41 (1964).
85. I. V. Solov'era, L. V. Voronova & A. N. Mal'tsev, *Russ. J. Phys. Chem.*, **48**, 298 (1974).
86. A. Slackza, *Int. Chem. Eng. Proc. Ind.*, **9**, 63 (1964).
87. E. Kowalska & M. Dziegielewska, *Ultrasonics*, **73**, (1976).
88. B. N. Tyutyunnikov & I. T. Novitskaya, *Maslob. Zhir. Prom.*, **25**, 13 (1959).
89. K. Ivanov, T. Popov & S. Slavov, *Ivz. Khim.*, **20**, 201 (1987).
90. T. Popov, D. G. Kissvrski, K. Ivanov & J. Pesheva, *Stud. Surf. Sci.* (Catal.), **31**, 191 (1987).
91. R. Ranganathan, N. N. Bakhshi & J. F. Mathews, *J. Catalysis*, **21**, 186 (1971).
92. R. Ranganathan, I. Mathur, N. N. Bakhshi & J. F. Mathews, *Ind. Eng. Prod. Res. Develop.*, **12**, 155 (1973).
93. E. Kowalska & M. Miszczyszyn, *Rocz. Chem.*, **46**, 233 (1972).
94. E. Kowalska, W. Kawalski & A. Slaczka, *Rocz. Chem.*, **39**, 1491 (1965).
95. V. I. Shekhobalova & L. V. Voronova, *Vestn. Mosk. Univ. Ser. Khim.*, **27**, 327 (1986).
96. K. S. Suslick & D. J. Casadonte, *J. Am. Chem. Soc.*, **109**, 3459 (1987).
97. E. A. Cioffi & J. H. Prestegarde, *Tetrahedron Lett.*, **27**, 415 (1986).
98. E. A. Cioffi, W. S. Willis & L. S. Svit, *Langmuir*, **4**, 692 (1988).
99. T. S. Mertes, US Patent 2 968 652 (1961).
100. *Jpn. Kokai Tokkyo Koho* Jap. Patent 59; 174 601 (1984).
101. F. Radenkov, K. Khristov, R. Kircheva & L. Petkov, *Khim. Ind. (Sofia)* **49**, 11 (1977).
102. C. A. Graves, D. F. Steiner & F. C. Hurdler, US Patent 3 231 513 (1966).
103. A. V. Romenskii, A. Y. Loboiko & V. I. Astroshenko, *Khim. Tekchnol.*, 39 (1986).
104. A. V. Romenskii, I. V. Popik, A. Y. Loboiko & V. I. Astroshenko, *Khim. Tekchnol.*, 21 (1985).
105. H. H. Clarke, R. Ranganathan, US Patent 4 086 184 (1978).
106. *Jpn Kok Tokkyo Koho* Jap. Patent 58 186 445 (1984).
107. W. Lintner & D. Hanesian, *Ultrasonics*, **15**, 21 (1977).
108. H. B. Weiner & P. W. Young, *J. Appl. Chem.*, **8**, 336 (1958).

109. B. Cornils, H. Bahrmann, D. E. Hamminkeln, W. Lipps & W. Konkol, German Patent DE 3511428 Al (1988).
110. A. N. Maltsev, *Russ. J. Phys. Chem.*, **50**, 995, (1976).
111. K. J. Moulton, S. Koritala, E. N. Frankel, *J. Am. Oil Chem. Soc.*, **60**, 1257 (1983).
112. K. J. Moulton, S. Koritala, K. Warner, E. N. Frankel, *J. Am. Oil Chem. Soc.*, **64**, 542 (1987).
113. P. Boudjouk & B. H. Han, *J. Catal.*, **79**, 247 (1985).
114. B. H. Han & D. H. Sin, *Bul. Kor. Chem. Soc.*, **6**, 247 (1985).
115. C. Petrier & J.-L. Luche, *Tetrahedron Lett.*, **28**, 2347 (1987).
116. B. H. Han & P. Boudjouk, *Tetrahedron Lett.*, **22**, 2757 (1981).
117. B. H. Han & P. Boudjouk, *Organometallics*, **2**, 769 (1983).
118. A. Fuentes, J. M. Marunas & J. V. Sinesterra, *Tetrahedron Lett.*, **28**, 2947 (1987).
119. A. Fuentes, J. M. Marunas & J. V. Sinesterra, *Tetrahedron Lett.*, **28**, 2951 (1987).
120. A. Fuentes, J. M. Marunas & J. V. Sinesterra, *J. Org. Chem.*, **52**, 3875 (1987).
121. K. S. Suslick, *High Energy Processes in Organometal Chemistry*, ACS Symposium Series No. 333, 1987, p. 191.

3 Ultrasonically assisted organic synthesis

R. S. Davidson
City University, London, UK

3.1 Metal-catalysed reactions

It does not seem so many years ago that organic chemists used to prepare their own solutions of alkyl lithiums by a finger-skinning process of cutting up hammered out sheets of lithium. An appreciation of the power of ultrasound to clean and etch surfaces of the alkali metals and to create dispersions has lifted us from the early days of the industrial revolution.

Similarly, a few years ago, the handling of stubborn Grignard reagent-forming reactions was limited to either adding a crystal of iodine or a few drops of 1,2-dibromoethane in the hope of inducing reactions. The effectiveness of ultrasound in cleaning the surface of metals offers another, and well-proven route, for persuading these reluctant reactions to take place. Exploration of the parameters which control the cleaning of metal surfaces and in changing particle size has had an important beneficial effect upon metal-catalysed reactions. This section reviews some of the benefits but there still remain many unexplored territories for those with enquiring minds, good powers of observation and chemical manipulative skills.

3.1.1 Group I metals

The two main features of this group of metals are their reductive power and their ability to undergo halogen exchange with the formation of metal alkyls.

The reducing power of the alkali metals has been used to produce radical anions of aromatic hydrocarbons and aromatic heterocyclic compounds and this process is aided by ultrasound[1-3] particularly when either tetrahydrofuran or 1,2-dimethoxyethane is used as a solvent. Metal ketyls can be used for a variety of purposes, which include the drying of solvents, and alkali metals for dehydrodimerization (e.g. preparation of bipyridyl from pyridine) and for transformation of aromatic into alicyclic compounds (Birch reduction[4]).

Anthracene can be reduced to its dianion by lithium under the influence of ultrasound. Addition of water leads to the production of 9,10-dihydroanthracene.[5] Diphenylacetylene gives its radical anion under similar conditions, and this dimerizes. The dianion can be trapped with appropriate species, as is shown in Scheme 1.

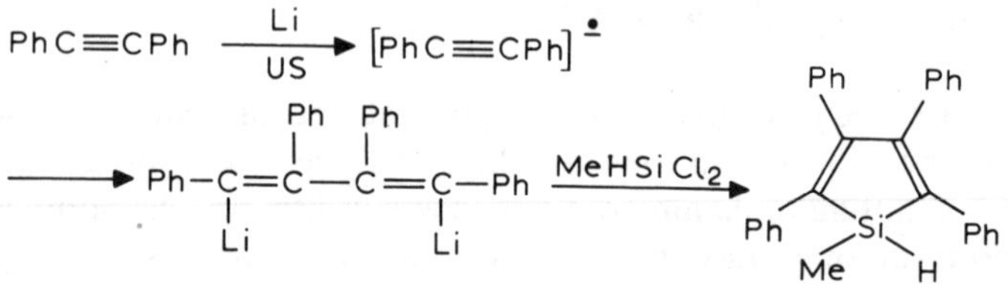

Scheme 1

The reduction of many carbonyl compounds by alkali metals has been shown by ESR spectroscopy to give radical anions which are often the precursors of reduction products. The ultrasonically aided reduction of (+)-camphor by lithium, sodium or potassium in tetrahydrofuran has been shown to give the same distribution of products as when liquid ammonia is used as solvent and ultrasound is not applied,[6] suggesting that ketyl radicals are intermediate in these reactions.

The ability of ultrasound to promote ketyl radical formation has been used in some elegant ring-closure reactions.[7] The yields of the products increased as the steric size of the alkyl groups (R^1, R^2) on the nitrogen were increased.

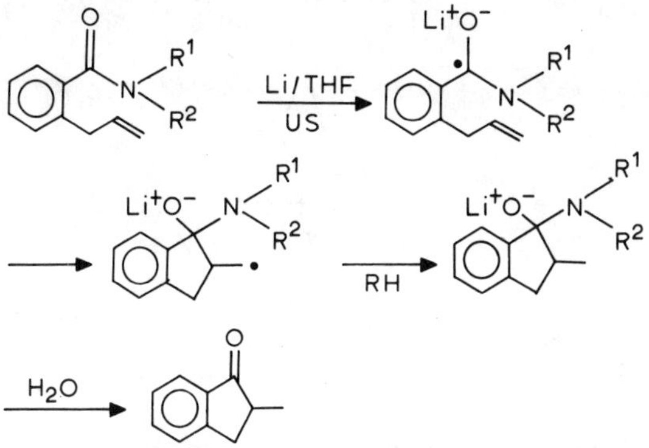

The formation of alkyl lithiums by reaction of bromoalkanes with lithium is a reaction of great importance in synthetic chemistry and is facilitated by the application of ultrasound. This is one of the reactions where a humble laboratory ultrasonic cleaning bath can work marvels.[8] Not surprisingly, lithiation is more effective with bromides than chlorides and the reactivity of the bromides is primary > secondary > tertiary. Methodology has now been developed so that alkyl lithiums can be generated *in situ*[9] and this promises to be of industrial value. The greater reactivity of bromoalkanes compared with chloroalkanes has been used to selectively activate a bifunctional molecule.

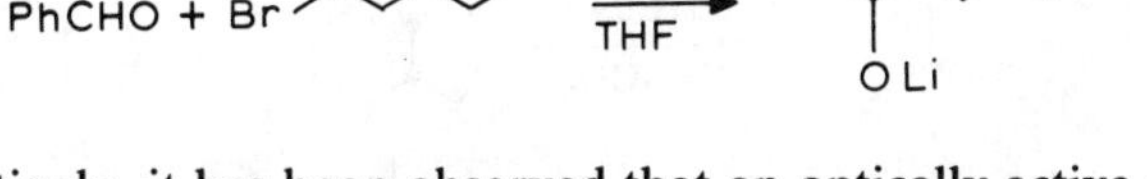

Interestingly, it has been observed that an optically active secondary bromoalkane can be lithiated and the resultant species reacted with a ketone to give a tertiary alcohol with a high degree of retention of configuration at the chiral atom provided that the reaction temperature is kept low ($-50°C$).[10] Since racemization normally occurs in such reactions, it was proposed as an explanation of the observed retention of configuration, that the lithium caused cleavage of the carbon–halogen bond by dissociative electron capture, giving the alkyl lithium species with retention of configuration.

Product with inversion

Product with retention of configuration

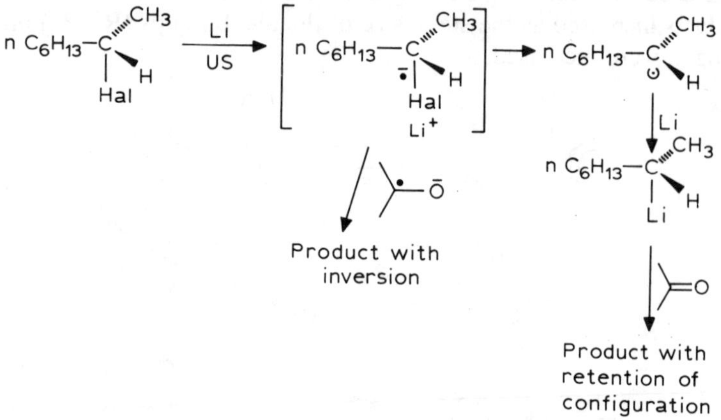

By the use of appropriate stoichiometric amount of reagents, lithium will induce the coupling of bromoalkanes to give alkanes[11] e.g.

$$2CH_3(CH_2)_3Br \xrightarrow[US]{Li,ether} CH_3(CH_2)_6CH_3$$

There are several other valuable features of ultrasonically aided lithiation reactions. Remarkably many of these reactions can be carried out in unpurified and wet solvents such as tetrahydrofuran.[8] Also vinylic, benzylic and allylic lithium species can be formed by this process, as can aryl lithiums.[11] Ultrasonically aided alkyl lithium formation has not only been of value in synthetic sequences where such species are reacted with

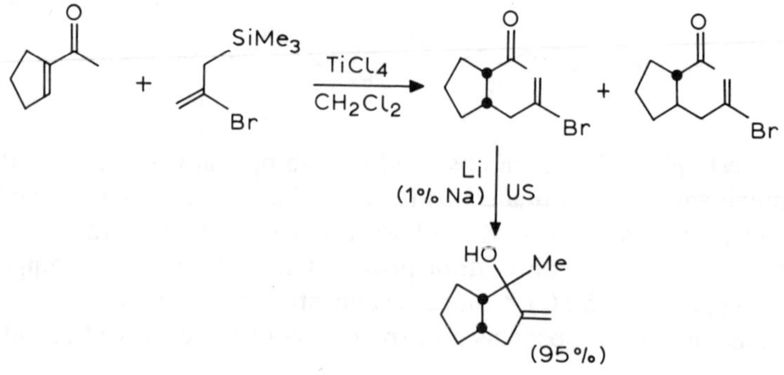

(95%)

simple carbonyl compounds[12] but also in the Bouveault reaction which is so valuable for the preparation of aldehydes.[13–15]

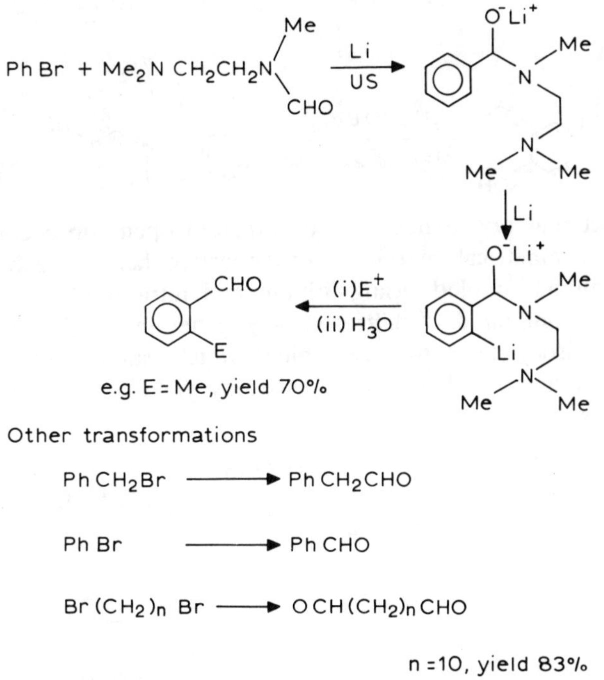

e.g. E = Me, yield 70%

Other transformations

$$Ph\,CH_2Br \longrightarrow Ph\,CH_2CHO$$

$$Ph\,Br \longrightarrow Ph\,CHO$$

$$Br\,(CH_2)_n\,Br \longrightarrow OCH(CH_2)_n\,CHO$$

n = 10, yield 83%

The aminoalkyl formamide is particularly useful because the tertiary amino group aids lithiation *ortho* to the latent aldehyde group. By use of this strategy, i.e. a double Bouveault reaction. *O*-Phthalaldehyde was prepared in 62% yield. The reactions are found to proceed with the greatest efficiency when tetrahydropyran as a solvent and radiation of high frequency (500 kHz—less cavitation) is used, but this is not the case when substantially lower frequencies, e.g. 50 kHz, are used. Since, albeit under special conditions, lithiation of aromatics can occur, it is not surprising that benzylic compounds can be lithiated.[8,17] The use of lithiated diamines to aid orthometallation of aromatic aldehydes has been put to good use.[16] By use of this method, aromatic carboxaldehydes are *ortho* hydroxylated. The products are thus produced by a route the direction of which is directly opposite to commonly used methods, e.g. the Reimer–Tiemann reaction.

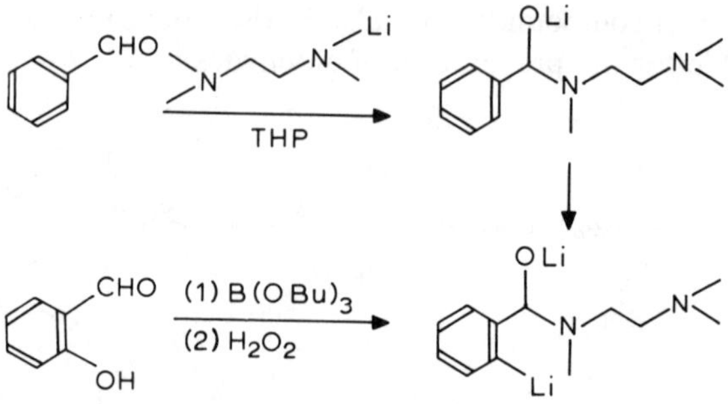

The fact that aryl halides can be lithiated opens up a convenient route to symmetrical biaryls.[17] More-reactive halides such as acyl halides can be coupled, using lithium and ultrasound, to give 1,2-dicarbonyl compounds.[17] Ultrasonically generated aryl sodium will react with isocyanates to give amides which can be ortholithiated under the influence of ultrasound, thereby yielding 1,2-disubstituted compounds.[18]

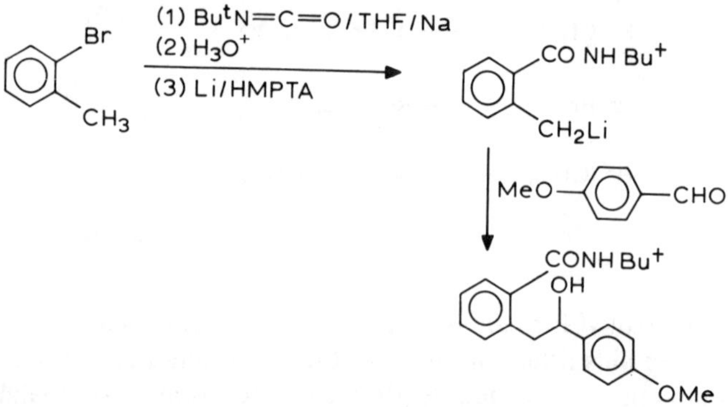

The formation of the alkyl lithiums under ultrasonic conditions has been used to generate a variety of carbanions and ylides:[19]

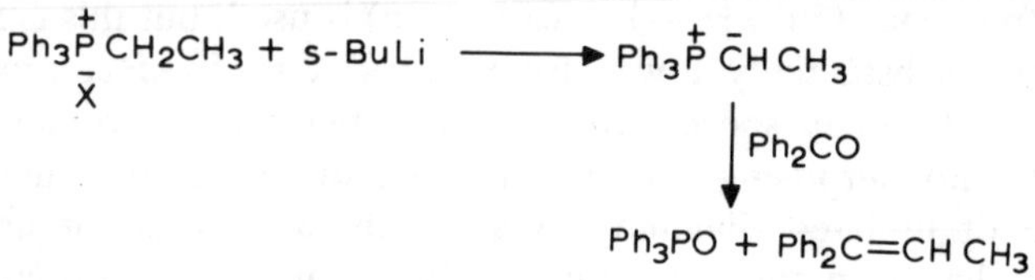

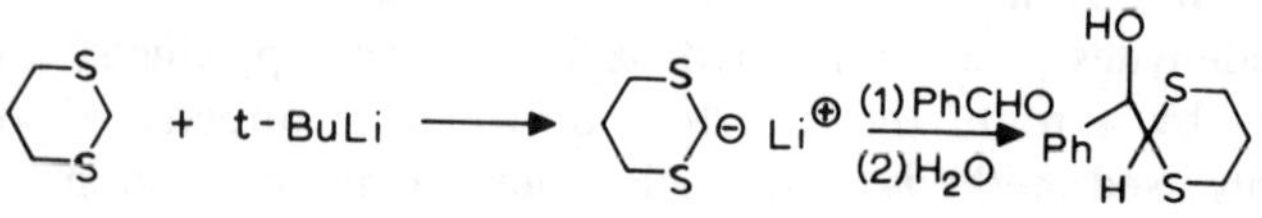

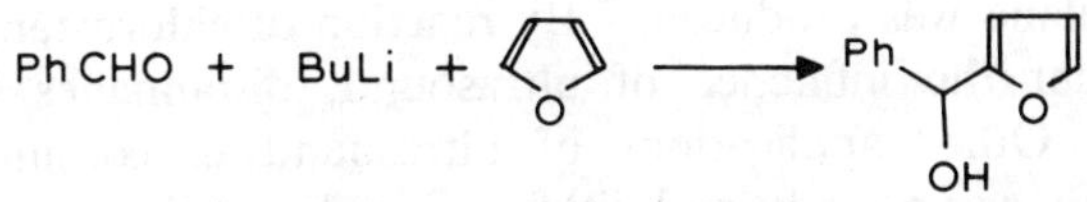

PhCHO + BuLi + (furan) ⟶ Ph—⟨⟩ with OH

Of particular use is that of ultrasonically activated lithium to prepare to lithium amides from hindered amines, e.g.

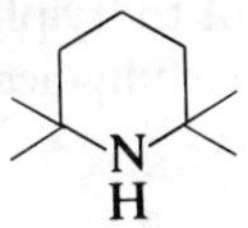

Pr_2^iNH

Methodology has been developed so that lithium diisopropylamide can be generated *in situ* using a mixture of lithium, an alkyl halide and di-isopropylamine.[19]

Examples of ultrasonically aided Wurtz reactions are not restricted to the formation of carbon–carbon bonds. Of particular interest, in view of their potential application in resist technology, are polysilanes[20,21] and examples of forming silicon–silicon–carbon bonds from silicon halides are well documented.[22,23]

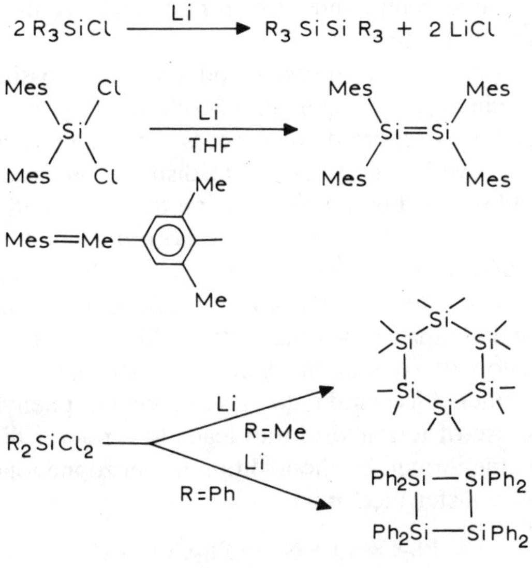

$2 R_3SiCl \xrightarrow{Li} R_3SiSiR_3 + 2 LiCl$

Mes–Si(Cl)(Cl)–Mes $\xrightarrow[THF]{Li}$ Mes(Mes)Si=Si(Mes)Mes

Mes=Me

R_2SiCl_2 $\xrightarrow[R=Me]{Li}$ (cyclohexasilane)

$\xrightarrow[R=Ph]{Li}$ Ph₂Si—SiPh₂ / Ph₂Si—SiPh₂

By use of dichlorosilanes a variety of products have been obtained.[23,24]

It should, however, be noted that other workers attempting to repeat the reaction of dimesitylsilyldichloride with lithium failed to obtain the product obtained by Boudjouk *et al.* but found instead that

R. S. Davidson

a cyclic trisilane was produced.[25] By reaction of chlorostannanes with lithium under the influence of ultrasound, distannanes have been produced.[23] Other applications of ultrasound in conjunction with organosilicon compounds and lithium has been the preparation of tri(phenyldimethylsilyl) methane.[26]

The reductive power of lithium is well illustrated by its capability to split the carbon–phosphorus bonds of triarylphosphines to give lithium phosphides[27,28] and the reaction is usefully accelerated by the application of ultrasound.

$$Ph_3P \xrightarrow[US]{Li} Ph_2\overset{\ominus}{P} \; Li^+$$

$$Ph_2P(CH_2)_n \; P \, Ph_2 \xrightarrow[US]{Li} Ph \; \overset{\ominus}{\underset{Li}{P}}(CH_2)_n \; \overset{\ominus}{\underset{Li}{P}}Ph$$

$$\xrightarrow{RHal} Ph \; \underset{R}{P}(CH_2)_n \; \underset{R}{P} \, Ph$$

The reaction can be used to prepare 1,2-bis(diphenylphosphino)-ethanes and related compounds, which find extensive use as ligands in organometallic chemistry.

The use of ultrasound to disperse sodium and potassium has found use in organic chemistry. Dispersion of sodium to aid the formation of phenyl sodium[2] was reported some years ago. Remarkably, use of a cleaning bath proved to be sufficient to disperse potassium in toluene and xylene and so good was the dispersion that the solution developed a silvery blue colour.[29] Sodium proved more difficult to disperse and xylene was found to be the only suitable solvent for this purpose. The highly dispersed alkaline metals proved useful in catalysing Dieckman cyclizations and olefination via the Wittig–Horner route.

The dispersion of sodium in xylene is particularly useful for the cleavage of phenyldiselenide to give sodium phenyl selenide.[30] Although the use of tetrahydrofuran leads to a poorer dispersion, the effect can be ameliorated by the addition of benzophenone, which acts as an electron-transfer mediator:

$$Ph_2C{=}O + Na \rightarrow Ph_2\dot{C}O^-Na^+$$

$$Ph_2\dot{C}O^-Na^+ + PhSeScPh \rightarrow PhS\dot{e} + NaSePh$$

Another advantage of using benzophenone is that the reaction becomes self-indicating, i.e. when all the sodium has been utilized a colour due to the ketyl radical is discharged. A dispersion of sodium in refluxing tetrahydrofuran has found use in the synthesis of polycyclic alkenes in which the crucial step is reductive decyanation.[31]

The cyclic sulphones undergo ring opening when subjected to ultrasound in the presence of potassium and an alkylating agent.[32]

The distribution of the products obtained with the unsymmetrically substituted sulphone suggests that radical anions are intermediates. A number of di- and trisubstituted sulphol-3-enes undergo cheletropic extrusion of sulphur dioxide on reaction with ultrasonically dispersed potassium.[33] The authors ruled out the possibility that these reactions were single electron-transfer reactions. In the case of sulphol-2-enes, which also undergo ring opening in the presence of potassium, they suggested that the alkyl metal first reduces the carbon–carbon double bond.

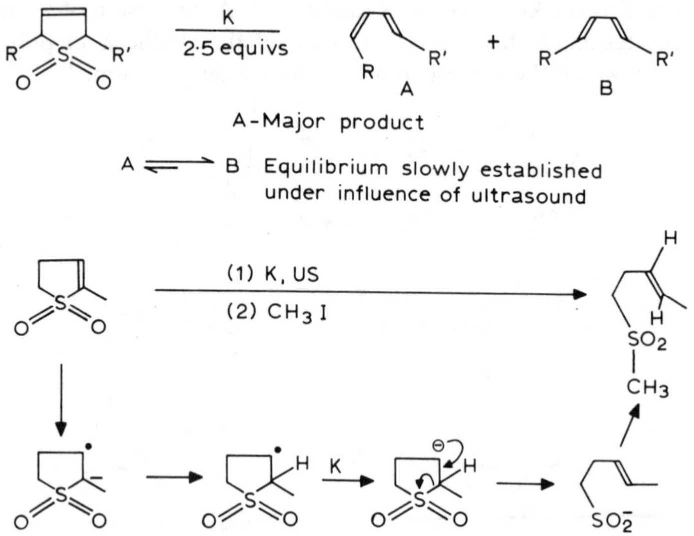

A—Major product

A $\rightleftharpoons$ B Equilibrium slowly established
under influence of ultrasound

Sodium in tetrahydrofuran under the influence of ultrasound has been shown to promote reaction between aryl halides and isocyanates, which ultimately leads to the introduction of a carboxamide group into the aromatic ring.[18] This reaction presumably occurs by an aryl sodium intermediate.

A sodium–potassium alloy has been sonicated in the presence of *n*-hexyltrichlorosilane in pentane to give a network polymer[34] the identity of which has been carefully established. Such new materials offer great promise as resists (they undergo cross-linking upon irradiation), etc. in the growing electronics industry.

Sodium hydride, which is used as a strong base for the formation of a carbanion from dimethyl sulphoxide, can also be activated by ultrasound.[35]

3.1.2 Group II Metals

For historical reasons it is worthwhile beginning this section with a consideration of the chemistry of magnesium, since it was just over three decades ago that ultrasound was found to enhance the reactivity of magnesium towards alkyl bromides and that the so-formed Grignard reagents reacted efficiently with aluminium.[36] The use of ultrasound to aid the formation of Grignard reactions has become widespread following a thorough study of the scope of the reaction.[8,9] Remarkably, such is the efficiency of ultrasound for cleaning the surface of magnesium that its application facilitates the formation of Grignard reagents in wet ether. More recently, the effects of adding known amounts of water to ether upon the induction period for the formation of butan-2-yl magnesium bromide has been studied[37] and even with 50% saturated ether the induction period was only 6–8 min.

Ultrasound has enabled allylic Grignard reagents to be prepared[38] as well as the more classical aryl magnesium halides as in ref 39. The magnesium generated by its reaction with anthracene in tetrahydrofuran is remarkably reactive and can be used to generate other reactive metals species which with appropriate ligands give isolatable metal complexes.[40]

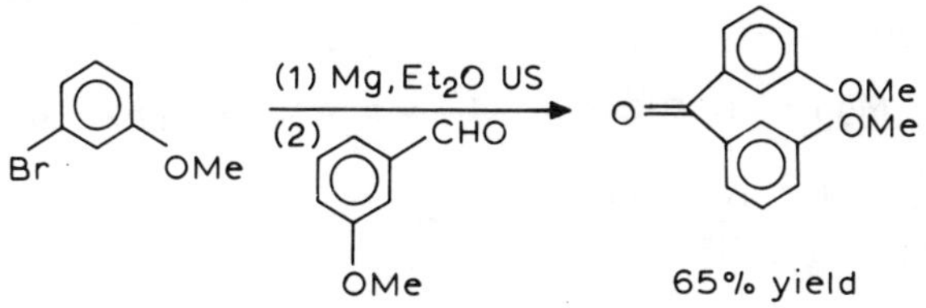

In some stereocontrolled Grignard reactions[41] the application of ultrasound has been found to improve product yield.

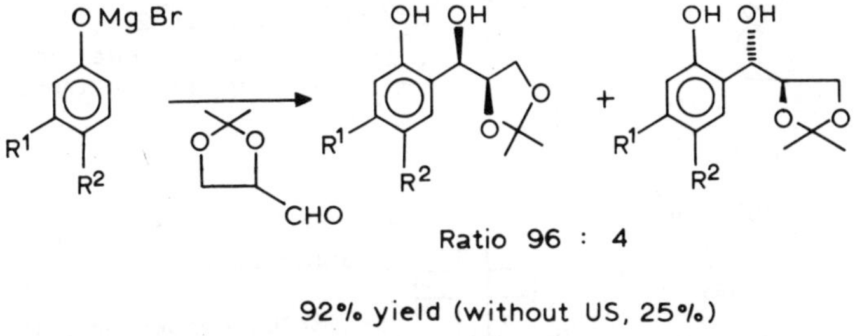

Ratio 96 : 4

92% yield (without US, 25%)

Perfluoroalkyl alcohols have been synthesized by the reaction of perfluoroaldehydes with in-situ generated alkyl or allyl Grignard reagents.[42]

$$CF_3CHO + RX \xrightarrow[\text{(2) } H_2O]{\text{(1) } Mg, Et_2O, US} CF_3RCHOH$$

The facile formation of Grignard reagents has been used in the preparation of isotopically labelled amphetamines via a substituted phenylboronic acid.[43]

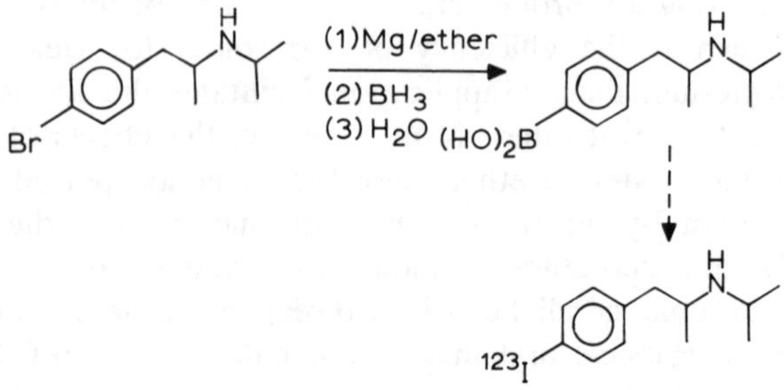
(1)Mg/ether
(2)BH3
(3)H2O
Br
(HO)2B
123I

The use of ultrasound has brought to fruition the originally conceived route for the preparation of triorganylboranes,[44] i.e. by reaction of Grignard reagents with trimethyl borate or boron trifluoride.

$$3\ RMgBr + B(OMe)_3 \xrightarrow{US} R_3B + 3\ MgBrOMe$$

$$3\ RX \xrightarrow[US]{Mg,\ BF_3,Et_2O} R_3B$$

$$\text{e.g. } PhBr \longrightarrow Ph_3B\ (97\%)$$

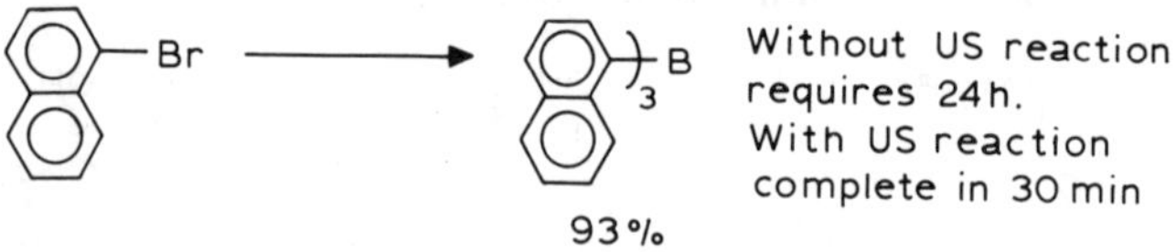
Br
B
3
93%
Without US reaction requires 24h.
With US reaction complete in 30 min

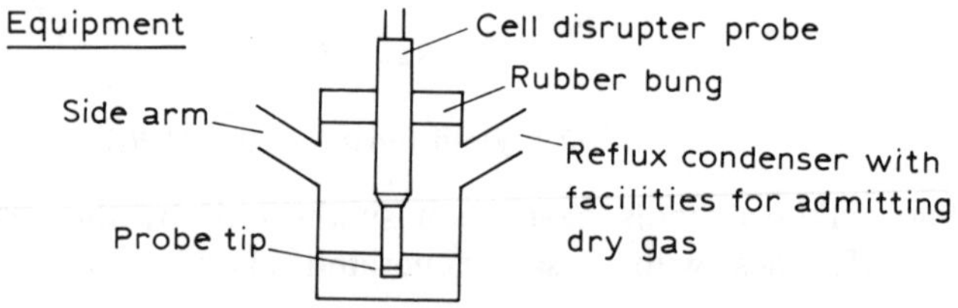
Equipment
Cell disrupter probe
Rubber bung
Side arm
Reflux condenser with facilities for admitting dry gas
Probe tip

Magnesium is also a reducing agent and is known to reduce carbonyl compounds by a one-electron process to give ketyl radicals which subsequently dimerize to give pinacols. Magnesium has also been shown to reduce anthracene in tetrahydrofuran to give its radical anion.[40] This species can act as an electron-transfer catalyst and promote the coupling of chlorosilanes and chlorostannanes.[23] The anthracene–magnesium complex has also been of value in the preparation of allyl Grignard reagents,[38] since in this way the concentration in the allyl Grignard reagents is kept low, thereby avoiding coupling with unreacted allylic halide.

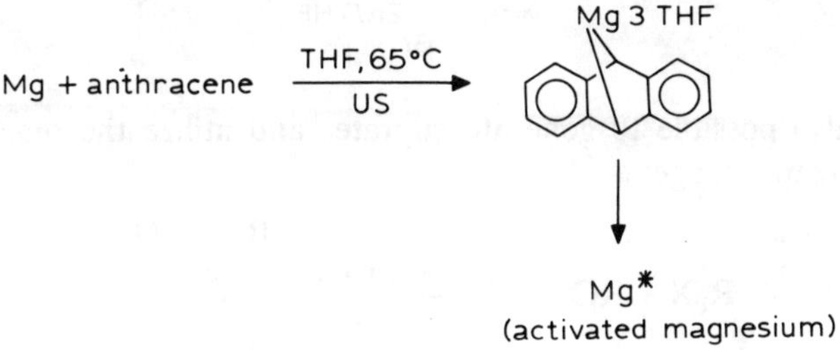

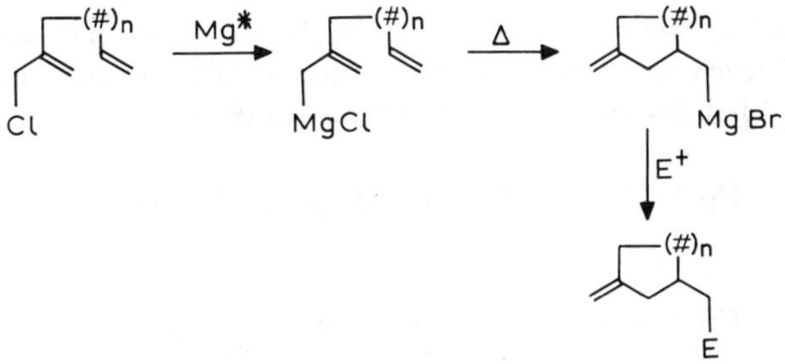

Some most spectacular and valuable results have been obtained by the ultrasonic activation of zinc. The soft metal is eroded by the ultrasound thereby continuously generating a reactive surface with the concomitant removal of products from the surface. Perfluoroalkylzincs can readily be formed and react with carbonyl compounds in the expected way to give alcohols.[45]

$$CF_3I \xrightarrow[\text{(2) } R_2CO,\text{ (3) } H_2O]{\text{(1) Zn/DMF}} CF_3CR_2 \underset{OH}{|}$$

Other reactions of perfluorozinc halides (prepared in either DMF or THF with a laboratory ultrasonic cleaner) also includes coupling reactions when an appropriate catalyst is present.[46]

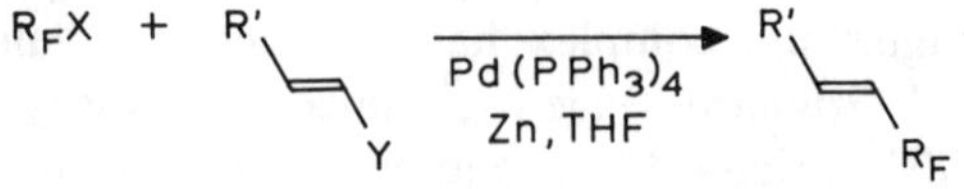

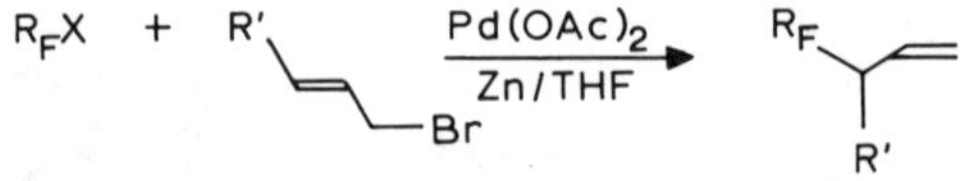

$$R_FX + R' \diagdown_Y \xrightarrow[\text{Zn, THF}]{\text{Pd(PPh}_3)_4} R' \diagdown_{R_F}$$

$$R_FX + R' \diagdown_{Br} \xrightarrow[\text{Zn/THF}]{\text{Pd(OAc)}_2} R_F \diagdown_{R'}$$

It is also possible to generate cuprates and utilize the reactivity of these versatile reagents.[46]

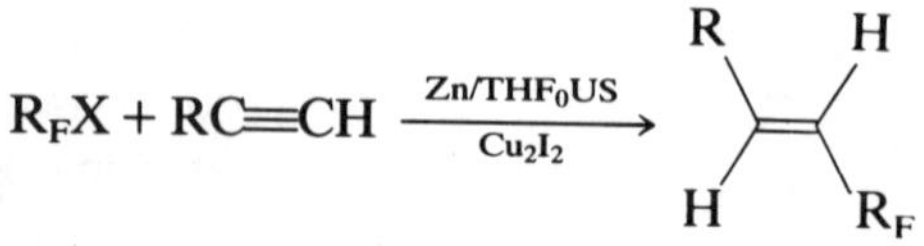

$$R_FX + RC\equiv CH \xrightarrow[\text{Cu}_2\text{I}_2]{\text{Zn/THF}_0\text{US}} \begin{array}{c} R \quad H \\ \diagup\diagdown \\ H \quad R_F \end{array}$$

A particularly clever use of ultrasonically activated zinc has been to prepare dicyclopentadienyl titanium for catalysing the reaction of fluoroalkyl halides with substituted buta-1,3-dienes.

$$Cp_2TiCl_2 \xrightarrow[\text{US}]{\text{Zn}} Cp_2Ti \quad (Cp = \text{cyclopentadienyl})$$

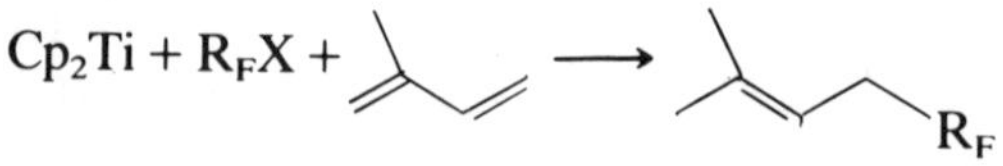

$$Cp_2Ti + R_FX + \diagup\diagdown\diagup \longrightarrow \diagdown\diagup\diagdown_{R_F}$$

Use of cuprous salts, palladium and nickel complexes have been found to catalyse the formation of the organozinc when the carbon–halogen bond to be activated is less reactive than that of the carbon–iodine bond.[47]

$$CF_3CCl_3 \xrightarrow[\text{Cu}_2\text{Cl}_2]{\underset{\text{DMF}}{\text{Zn}}} CF_3CCl_2ZnCl \xrightarrow{RCHO} \overset{\overset{\displaystyle OH}{|}}{RCHCCl_2CF_3}$$

Other catalysts (PdCl$_2$(PPh$_3$)$_2$ or NiCl$_2$(PPh$_3$)$_2$) have also been used. Reaction of chiral aldehydes (chromium tricarbonylcomplexes of aromatic aldehydes) with perfluoroalkylzincs has been shown to give alcohols with asymmetric induction the degree of which is dependent upon the structure of the perfluoroalkylzinc.[48]

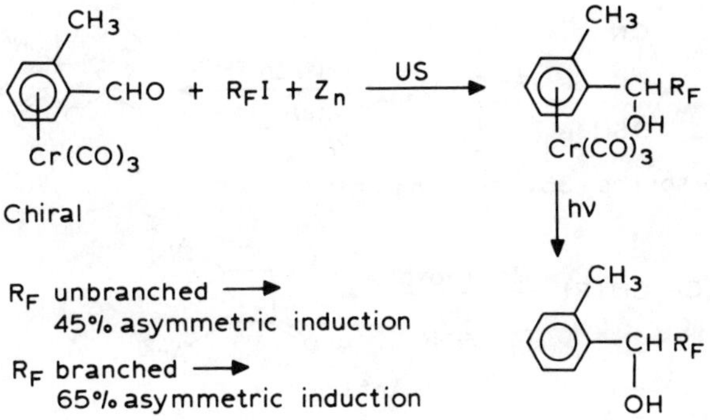

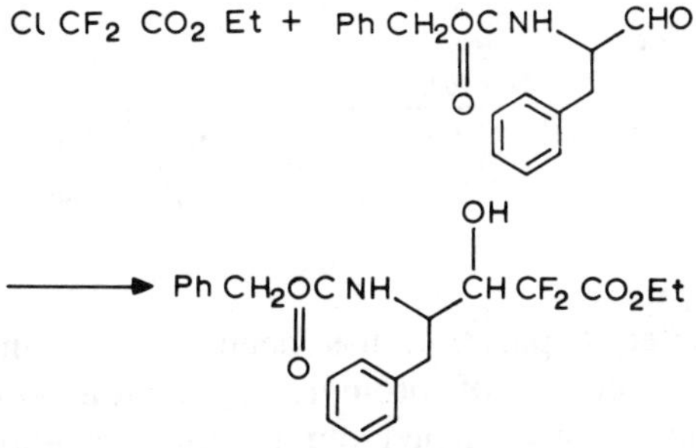

A classical use of zinc in organic chemistry is in the Reformatsky reaction and ultrasound promotes this reaction. A wide variety of classical Reformatsky reactions[49,50] have been carried out using ultrasound.

$$\text{RR'CO} + \text{BrCH}_2\text{CO}_2\text{Et} \xrightarrow[\text{US}]{\text{Zn dioxane}} \text{RR'CH(OH)CO}_2\text{Et}$$

Products were obtained using relatively short reaction times and in high yield. Perfluoroalkylaldehydes have also been reacted with ultrasonically generated organozinc compounds.[51]

The synthesis of fluorinated β-keto-γ-butyrolactones has been accomplished via organozinc intermediates.[52] No reaction was observed in the absence of ultrasound.

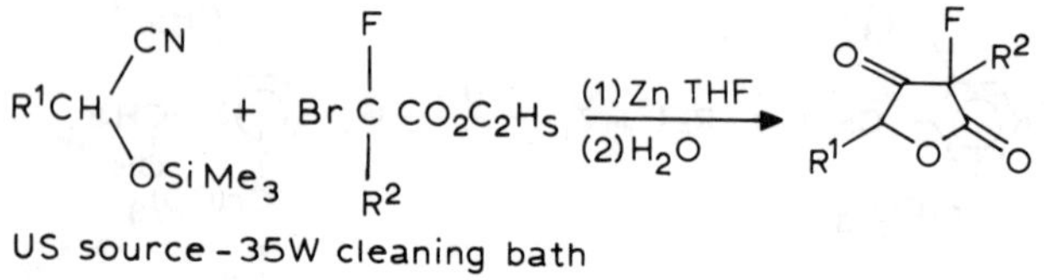

US source - 35W cleaning bath

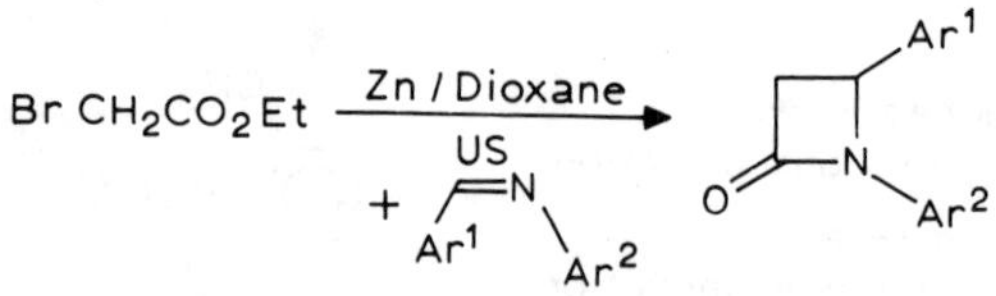

e.g.

Ar^1	Ar^2	Yield
C$_6$H$_4$Me(p)	C$_6$H$_4$OMe(p)	95%
C$_6$H$_4$Cl (p)	C$_6$H$_4$OMe(p)	77%

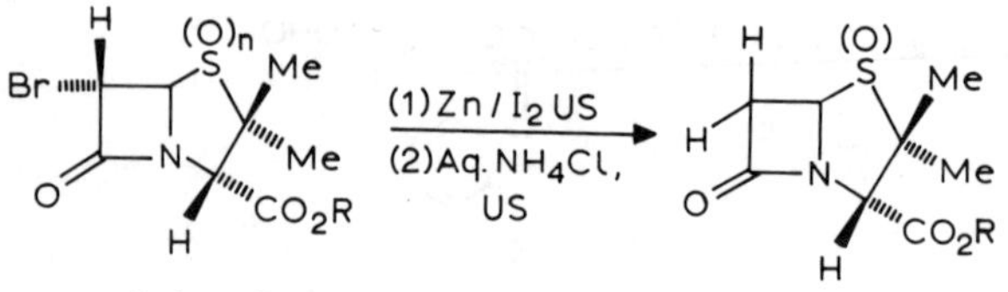

n = 0, 1 or 2

An interesting related application has been to the synthesis of
β-lactams.[53] The replacement of bromine in α-bromoketones by
hydrogen can be conveniently carried out by the formation of the
organozinc compound followed by hydrolysis.[54] This provides a useful
alternative to removal of bromine by the use of tributyltin hydride.
Allylzincs can be prepared in tetrahydrofuran by the application of
ultrasound and the organozincs can be shown to add to alkynes.[55]

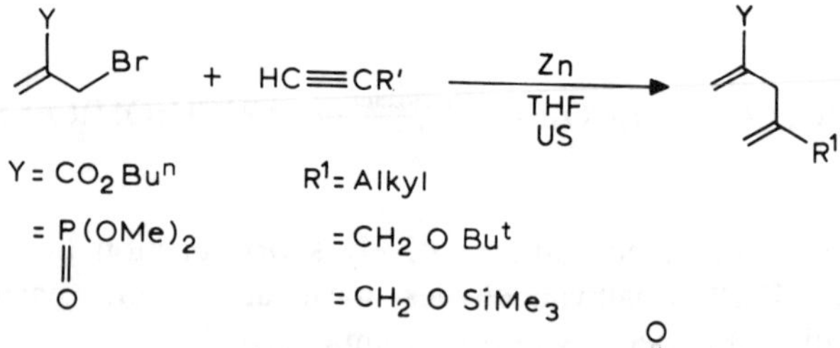

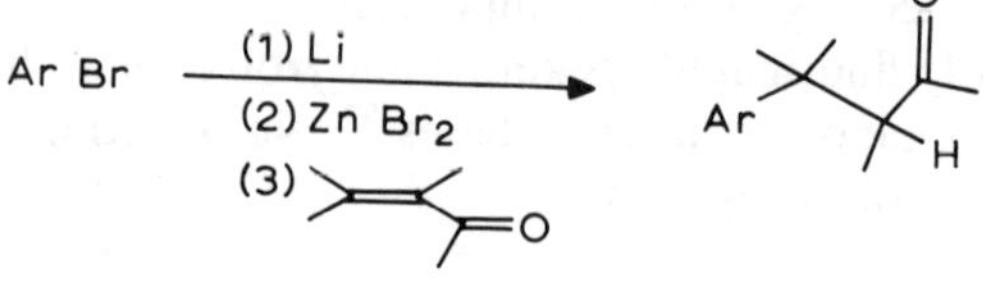

Y = CO$_2$Bu^n R^1= Alkyl

  = P(OMe)$_2$ = CH$_2$O Bu^t
    ‖
    O   = CH$_2$O SiMe$_3$

Amazingly, allylic halides react with zinc in aqueous solution to give products which react with aldehydes to give alcohols.[56]

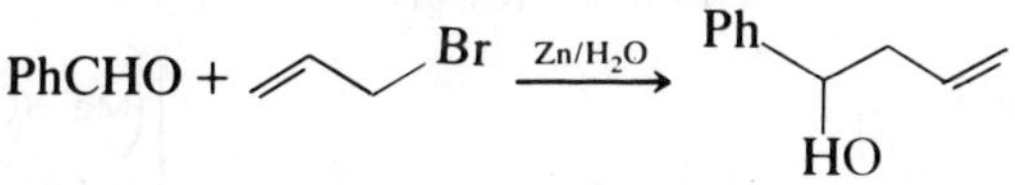

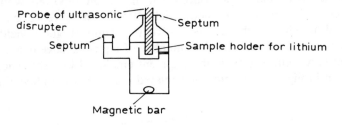

$$PhCHO + \text{~~~}Br \xrightarrow{\text{Zn/H}_2\text{O}} Ph\text{~~~}$$

It is thought that these reactions do not involve classical organozinc compounds but rather that the zinc initiates the reaction, which operates via a single electron-transfer mechanism. The finding that benzylic and tertiary butyl halides will not react under these conditions strongly suggests that simple alkylzincs are not involved. The reaction of diarylzincs with $\alpha:\beta$-unsaturated ketones in the presence of copper salts leads to conjugate addition,[57] the reactions being particularly mild. Details of the equipment have been given of[58] and the range of alkylzincs increased.

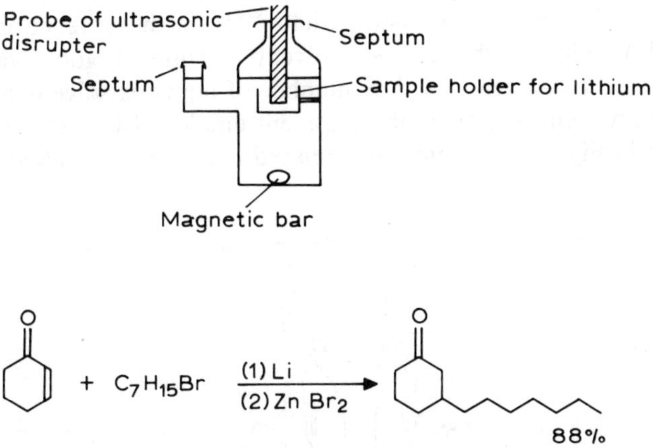

Probe of ultrasonic disrupter

Septum

Septum

Sample holder for lithium

Magnetic bar

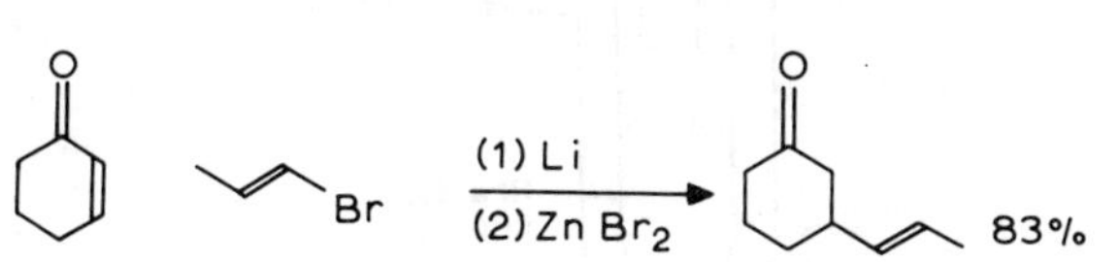

C$_7$H$_{15}$Br

(1) Li
(2) Zn Br$_2$

88%

(1) Li
(2) Zn Br$_2$

83%

These reactions have found use in synthesis, e.g. for $(\pm)$-β-cuparenone.[59]

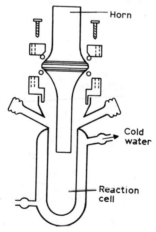

Conjugate addition of organozincs to α,β-unsaturated aldehydes in the presence of nickel acetylacetonate has been reported with the products being formed in good yield.[60,61] In these reactions it is essential to choose the correct solvent mixture if good yields of products are to be realized.[61] A number of aqueous solvent mixtures behave very well. The type of equipment employed in these reactions is shown in Fig. 3.1. A somewhat related reaction is the alkylation of

Fig. 3.1 An ultrasonic reaction vessel equipped with an ultrasonic horn.

vinylphosphine oxides by reaction with alkyl halides in the presence of a zinc–copper couple[62]

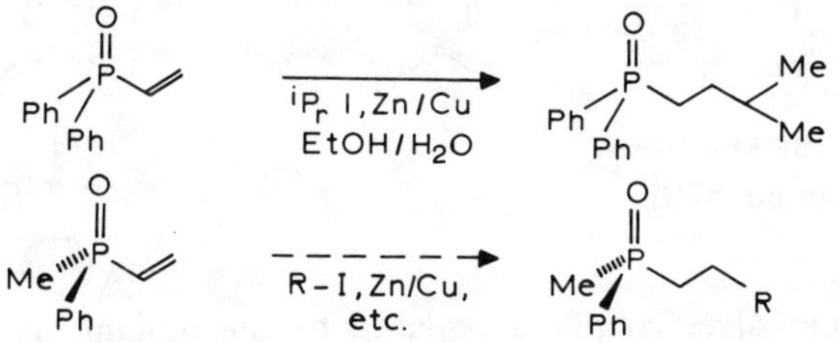

The conjugate addition of alkylzincs to $\alpha:\beta$-unsaturated ketones under the influence of cuprous iodide has also been accomplished in aqueous media.[63,64] This rather remarkable reaction is thought to be yet another single electron-transfer (SET) process in which the zinc reduces the alkyl halide to generate an alkyl radical which then adds to the β-carbon atom of the enone. Evidence in favour of the SET mechanism[64] comes from the finding that products typical of radical reactions, e.g. radical-induced annelation, can be formed and secondly from deuterium incorporation experiments.

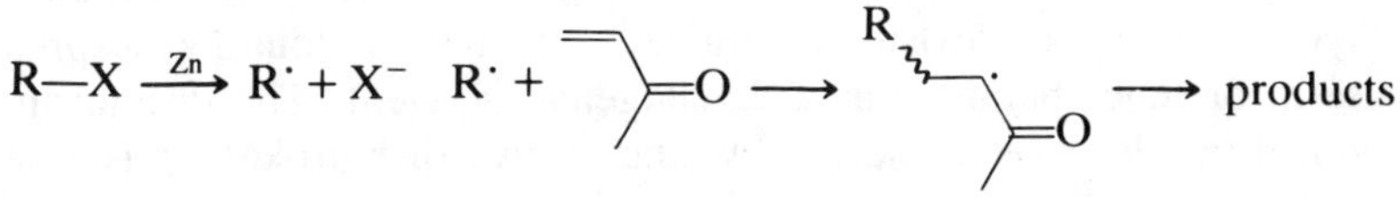

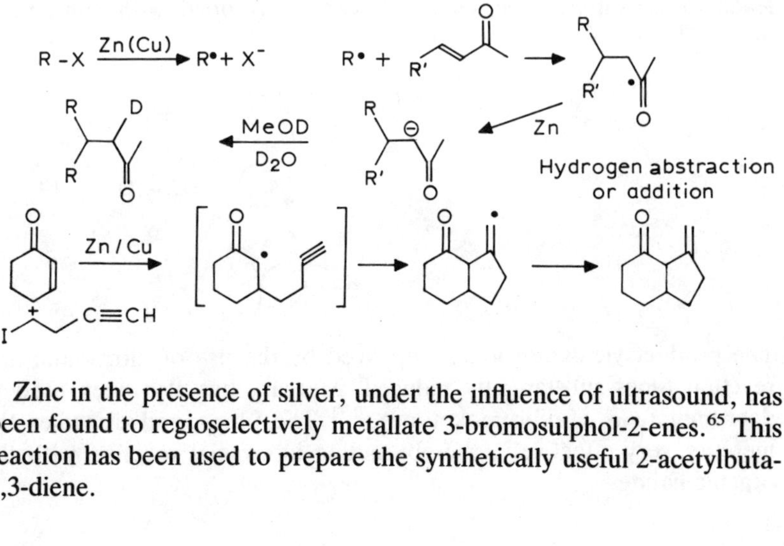

Zinc in the presence of silver, under the influence of ultrasound, has been found to regioselectively metallate 3-bromosulphol-2-enes.[65] This reaction has been used to prepare the synthetically useful 2-acetylbuta-1,3-diene.

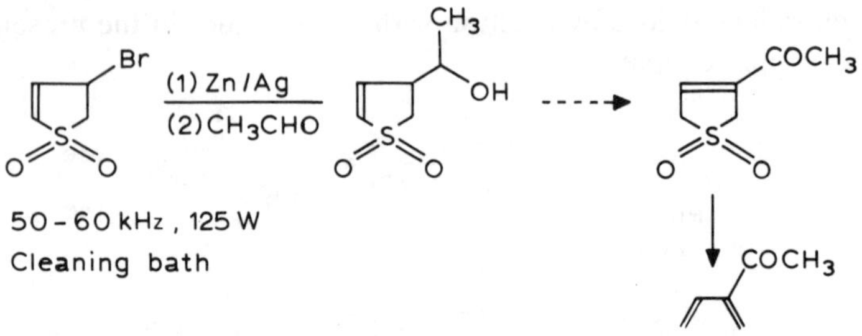

50 – 60 kHz , 125 W
Cleaning bath

If the zinc–silver couple is replaced by magnesium, alkylation proceeds to give the 2-alkylated sulphol-3-ene.

The use of zinc as a reducing agent is well known in organic chemistry. However, the use of zinc–nickel chloride (9:1) mixture in aqueous alcoholic solution with $\alpha:\beta$-unsaturated ketones leads to reduction of the double bond.[66] Alcohol can be replaced by hydrogen donors such as amines or even hydrogen itself. Several examples have been provided which testify to the fact that reduction of the carbon–carbon double bond is much faster than that of the carbonyl group.[66] Metallic zinc has been used to reduce benzo-1,4-quinone and anthra-9,10-quinone to the appropriate phenols which are then trapped by silylation with trimethylsilyl chloride.[67] Zinc has also found widespread use in organic chemistry as a dehalogenating agent. The dechloronation of trichloroacetyl choride by zinc to give dichloroketene is aided by ultrasound.[68] In the case of dibromoketones, debromination by zinc leads to 1,3-dipolar species which can be trapped with conjugated dienes.[69]

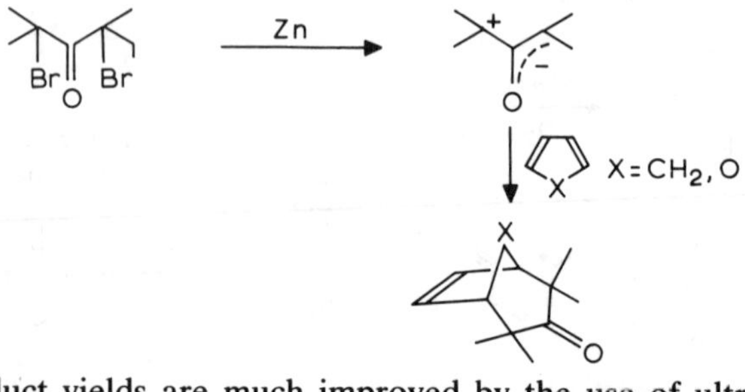

The product yields are much improved by the use of ultrasound and reaction times substantially reduced. Mercury has also been used to debrominate α,α'-dibromoketones.[70,71] The ultrasound disperses the mercury very effectively, thereby aiding interfacial reaction with the organic halide.

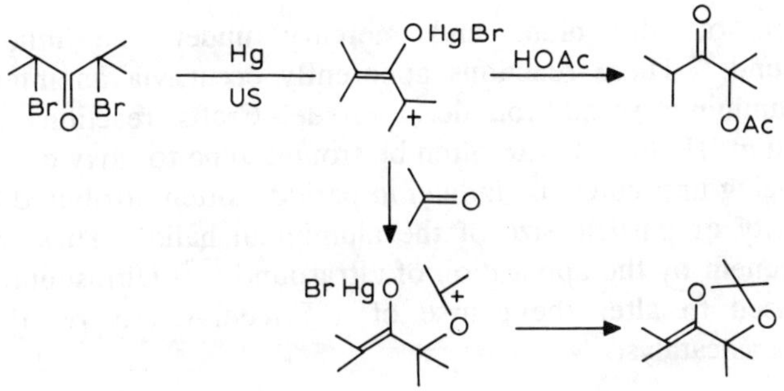

3.1.3 Elements of other groups

Reference has already been made to the usefulness of ultrasound for the preparation of trialkylboranes.[44] The process of hydroboration to accomplish anti-Markovnikov addition of the elements of water to olefins is well established. The hydroboration of some olefins presents problems because the intermediates generated in the reaction are insoluble in the reaction medium. This situation can be ameliorated by the use of ultrasound[72] and examples have been given which exemplify the beneficial use of this type of power.

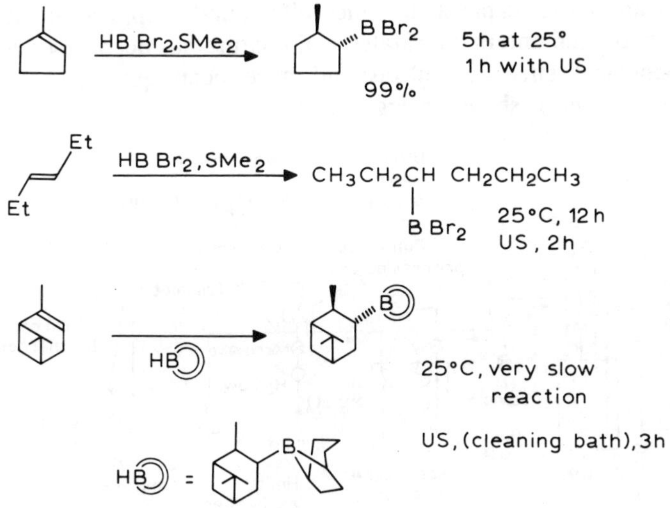

Trialkylaluminiums have been prepared from alkyl halides in the

86   *R. S. Davidson*

presence of aluminium and mercury under the influence of ultrasound.[73] These reactions apparently occur via an intermediate alkylaluminium sesquibromide.[74] Friedel–Crafts reactions involving aluminium III chloride can often be troublesome to carry out owing to there being unpredictable induction periods, often attributed to either the purity or particle size of the aluminium halide. These reactions often benefit by the application of ultrasound.[75,76] Ultrasound has also been used to alter the course of a Friedel–Crafts reaction using alumina as catalyst.[77]

Use of a variety of metals to catalyse the addition of hydrogen to carbon–carbon double bonds is not only a well-explored field but one of vast commercial importance. Chapter 2 contains a detailed description of the catalysts and the processes by which they are utilized. Suffice it to note that Raney nickel is activated by ultrasound and the relationship between power and frequency has been established for the hydrogenation of olive oil.[78] The hydrogenation of soya-bean oil using copper and nickel catalysts in which ultrasound is applied at the point where hydrogen enters the catalyst–oil mixture has been described and the beneficial effects of ultrasound have been quantified.[79,80] The equipment used is shown in Fig. 3.2.

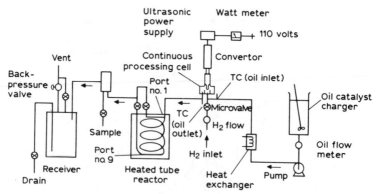

Fig. 3.2 Design of reactor for hydrogenation of soya bean oil assisted by ultrasound.[79]

Metals such as palladium and platinum can be used to effect dehydrogenation, e.g.

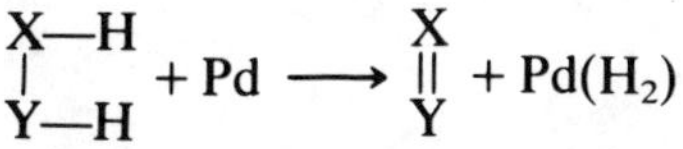

$$\begin{matrix} X\!-\!H \\ | \\ Y\!-\!H \end{matrix} + Pd \longrightarrow \begin{matrix} X \\ \| \\ Y \end{matrix} + Pd(H_2)$$

The hydrogen may be transferred to another species:

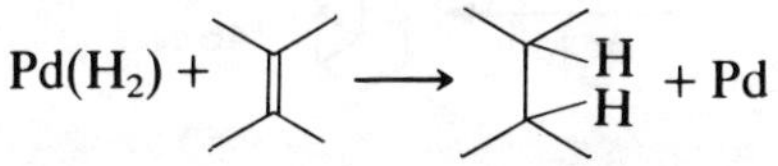

$$Pd(H_2) + \quad\longrightarrow\quad \begin{matrix} -H \\ -H \end{matrix} + Pd$$

Raney nickel has been used in this way to oxidize some glycophospholipids in the presence of deuterium and then to reduce the so-formed ketone with the incorporation of deuterium.[81] In this way deuterium was introduced regioselectively. Hydrazine and formic acid are easily dehydrogenated by palladized charcoal and this type of reagent mixture is very useful for the transfer of hydrogen to alkenes, dienes and vinyl ethers.[82] Another commercially important reaction is hydrosilylation:

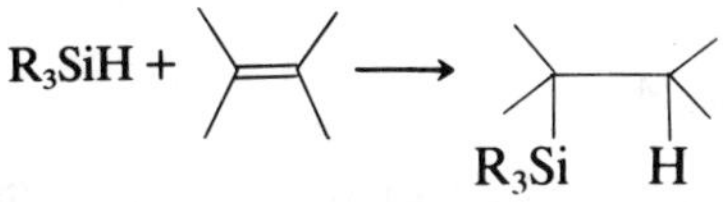

$$R_3SiH + \quad\longrightarrow\quad \underset{R_3Si \quad\quad H}{}$$

Platinum is the catalyst of choice for this reaction but nevertheless the reaction conditions are quite forcing. By use of ultrasound, the reactions can be carried out conveniently at room temperature.[83] It has also been reported that palladium (II) compounds will catalyse the hydration of alkynones.[84]

Examples of the use of ultrasonically activated copper and nickel in the Ullmann reaction have been reported.[85,86] Activated nickel has been found to promote the formation of biaryls from the triflate esters of phenols.[86]

3.2 Generation of reactive intermediates

Carbenes and carbenoids can be generated very simply using ultrasound. Although generation of carbenes by the reaction of *gem* dihalides with zinc/copper couples is known, the use of ultrasound to aid the reaction removes much of its capriciousness and leads to increased yields and shorter reaction times.[87,88]

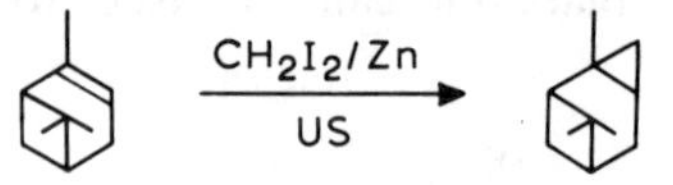

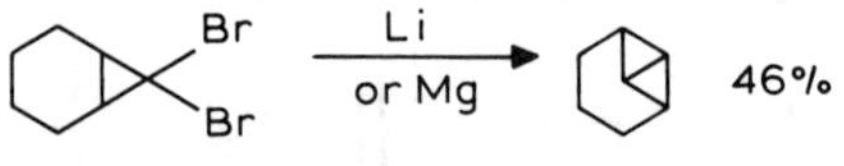

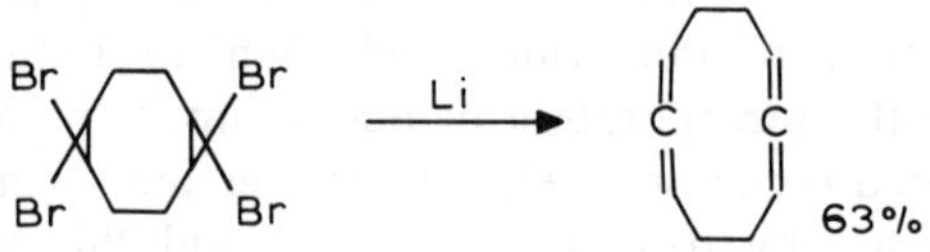

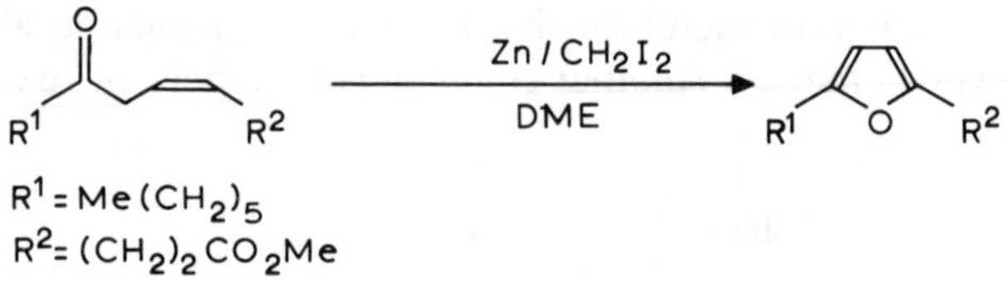

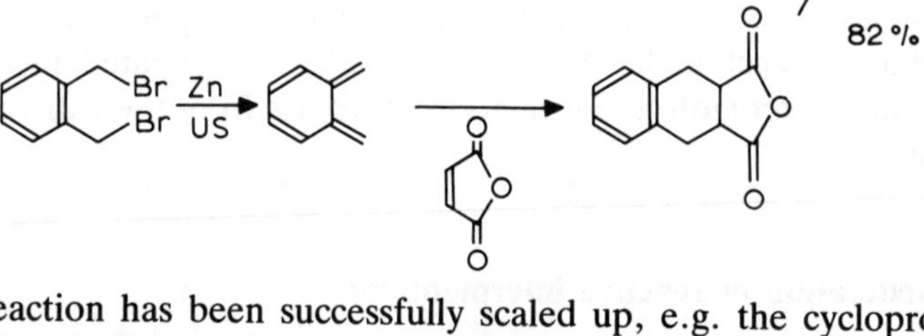

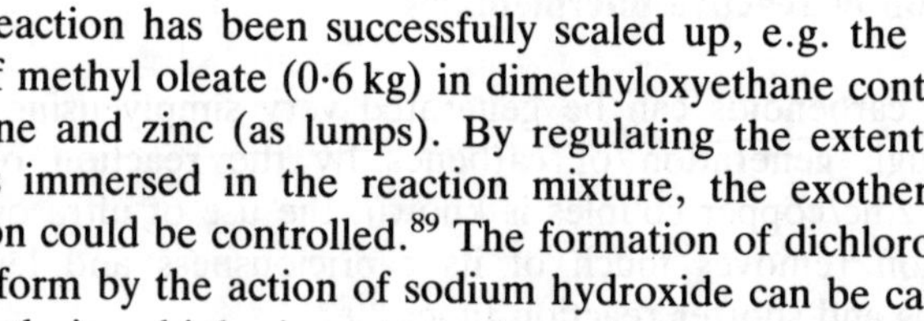

This reaction has been successfully scaled up, e.g. the cyclopropana-tion of methyl oleate (0·6 kg) in dimethyloxyethane containing diiodo-methane and zinc (as lumps). By regulating the extent to which the zinc is immersed in the reaction mixture, the exothermicity of the reaction could be controlled.[89] The formation of dichlorocarbene from chloroform by the action of sodium hydroxide can be carried out very efficiently in a biphasic system where the reaction mixture is agitated by ultrasound.[90] The use of powdered sodium hydroxide in chloroform solution is also a very good system.[90] Generation of a carbene in the presence of $\beta:\gamma$-unsaturated ketones has been shown to give furans.[91] A peculiar feature of this reaction is that the carbene is not incorporated into the product. The use of *gem* dihalides in which one halogen is less reactive than the other has been explored with fruitful results.[92] Thus, the reaction of bromochloromethane with lithium leads to displacement of the bromine and the alkyl lithium so produced will react with a number of carbonyl compounds. The initial product of the reaction is a chlorohydrin, which often undergoes further reaction to give an oxirane in high yield. Mention has already been made of the generation of silicon analogues of carbene (silylenes).[24] Other reactive species generated with the aid of ultrasound include *O*-quinonemethides (which often feature in the synthesis of steroids and anthracyclones). The treatment of 1,2-di(bromomethyl) benzene with zinc gives the appropriate quinone methide.[93]

Reactions in homogeneous solution

There are several clear examples of the application of ultrasound increasing the rate of reaction in homogenous solution, but the reasons for this increase are not always quite so clear. It has been observed that benzoyl azide decomposes to give nitrogen and the phenylisocyanate in benzene solution and that the rate of liberation of nitrogen is increased sevenfold by the application of ultrasound under the described experimental conditions.[94] Since the decomposition is a thermal process, there is the possibility that the increase in rate is due to a heating effect and presumably, not due to decomposition in the cavitation bubbles, since the acyl azide is relatively involatile.

The increased rate of ester hydrolysis (although not particularly marked) has been subjected to a variety of interpretations usually associated with cavitation.[95-98] An emerging fact, which seems to be true for most ultrasonically aided reactions is that the reaction rate decreases as the frequency of the ultrasound is increased. In a detailed study of the solvolysis of 2-chloro-2-methyl propane this has been shown to be the case.[97] By use of appropriate equipment, quite marked increases in the rate of reaction (20 times) were observed using 20 kHz ultrasound.[98] For this reaction it was found that varying the composition of the ethanol–water mixtures leads to a maximum rate enhancement at that composition for which the ultrasound is absorbed to the greatest extent. Under these circumstances solvent structure breaking is maximal and leads to a decrease in the extent of hydrogen bonding. Other useful reactions which occur in homogenous solution have been found to be accelerated by the application of ultrasound, but whether this is solely due to the increased efficiency of mixing is not clear. The protection of sugars via acetal formation with propan-2-one is one such reaction and of particular importance is the fact that the reaction can easily be scaled up.[99] With the great interest in the use of sugars as chiral synthons, this reaction is likely to see extensive use. The preparation of sydnones has in some case benefited by the application of ultrasound.[100]

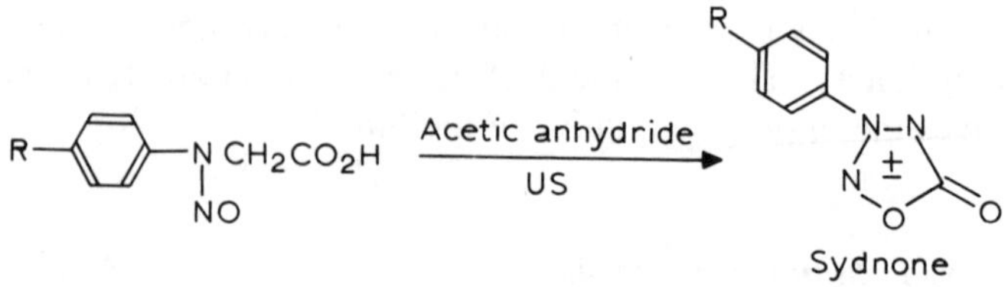

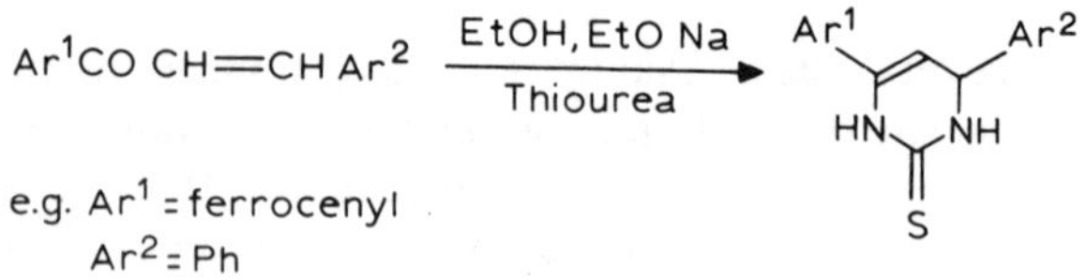

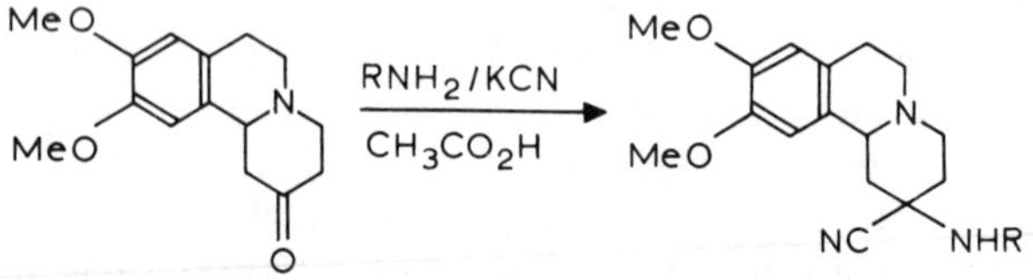

In those cases where the reaction is normally very sluggish, use of ultrasound usually reduces reaction time and increases product yields. Similarly those reactions between thiourea and chalcones which, because of substituent effects are normally somewhat slow can be speeded up by the use of ultrasound.[101] Examples have been reported of $(4 + 2)$ cycloaddition reactions of unsaturated nitrones to alkenes being accelerated by the use of ultrasound.[102] An example of the

Strecker reaction has been shown to benefit by the use of ultrasound in that application of energy in this way reduced reaction times.[103]

The hydrolysis of esters to give acids and alcohols by the action of alkalis has a rich history and is of undoubted commercial importance. Certainly the observation that ultrasound promotes the hydrolysis of sugar cane bagasse by hydrochloric acid containing lithium chloride is of note in this connection.[104] In accordance with theory, the rate of hydrolysis is increased as the frequency of the ultrasound is decreased from 40 to 25 kHz.

Although ultrasound is known to fragment macromolecules it also appears to affect the structure of polyelectrolytes, for example, polyacrylic acid and polysaccharides in solution. The binding of the terbium ion by such electrolytes appears, in general (poly(methyl methacrylate) is an exception), to be increased by ultrasound and as a consequence the quantum yield for light emission is increased.[105] Ultrasound has also been used to depolymerize calf thymus DNA.[106]

3.3 Ultrasonic mixing

3.3.1 Liquid–liquid reactions

Of particular importance in this field is the hydrolysis of esters. Early work reported that these reactions could be enhanced by the application of ultrasound.[107] In the cardinal paper on the subject it was reported that the hydrolysis of a sterically hindered 2,4-dimethylbenzoic acid esters by aqueous sodium hydroxide was dramatically increased by the application of ultrasound.[108] Some of the results are given in Table 3.1. As Table 3.1 shows, application of ultrasound caused a marked increase in rate and the improvement of

Table 3.1. Yields of 2,4-dimethylbenzoic acid produced by hydrolysis of its methyl ester

Reaction time	90 min	10 min	60 min
Conditions	Reflux[a]	US[a+b]	US[a+b]
	15%	15%	94%

[a] Mixture = lg ester + 10 ml 20% NaOH.
[b] Ultrasound frequency = 20 kHz.

yields of products meant that the separation of these from starting materials was much easier. In accordance with these findings the rate of hydrolysis of glycerides, e.g. glycerol tripalmitate, by sodium hydroxide is also increased by the application of ultrasound.[109] Such reactions also benefit through the use of phase-transfer catalysts. Use of lipophilic catalysts, e.g. tetra-n-heptylammonium bromide are very efficient because of their ability to transport the hydroxide ions into the ester layer. Many naturally occurring oils and waxes, e.g. rape oil, soya-bean oil and wool wax can be hydrolysed efficiently by aqueous alkali in conjunction with ultrasound and a phase-transfer catalyst.[110] Benefits of using this procedure include the facts that reaction times are shorter, the products possess a better colour, and reactions can be carried out at lower temperatures. By use of a whistle reactor[109] these reactions can be successfully scaled up. The function of a whistle reactor is to break up the wax particles, thereby producing very small droplets of oil. Thus, by increasing the surface area of the wax particles, mass transport of the phase-transfer catalyst is considerably enhanced.

The generation of carbanions and related species together with their participation in nucleophilic substitution reactions often benefits from the use of ultrasound. The C-alkylation of isoquinoline by the dimsyl anion has been reported.[111]

The autoxidation of 4-methyltoluene in the presence of poly(ethylene glycol) and oxygen gives both 4,4′-dinitrobibenzyl and 4,4′-dinitrostilbene.[112] However, when the reaction is carried out with ultrasonic agitation, 4-nitrobenzoic acid becomes the major product. Even with $para$-nitroethylbenzene the nitrobenzoic acid is obtained in reasonable yield. Presumably these reactions are involving the generation of the benzylic carbanion followed by its subsequent reaction with oxygen—either electron transfer giving a benzylic radical or nucleophilic attack upon oxygen. The ultrasonically aided reaction presumably occurs via benzylic hydroperoxides. Benzylic carbanion formation of isoquinoline Reissert compounds followed by alkylation is favoured by the application of ultrasound in conjunction with a phase-transfer catalyst.[113]

Nucleophilic substitution reactions carried out in organic–aqueous solution media with the aid of phase-transfer catalysts often proceed far more rapidly when ultrasound is applied. Not only does the ultrasound increase the efficiency of mixing but, as pointed out previously, it decreases the particle size of the organic phase. This was shown to be the case with the reaction of thiocyanate anions with alkyl halides.[114]

$$R - Br \xrightarrow[-CNS]{} RCNS + Br^-$$

Not surprisingly, lipophilic phase transfer catalysts proved to be the most effective. The use of ultrasound cannot, however, overcome the steric problems associated with nucleophilic substitution at secondary and tertiary carbon atoms. Indeed, in the formation of azides from alkyl halides by reaction with sodium azide only benzylic and allylic halides gave good yields, with primary halides reacting far less efficiently.[115]

3.3.2 Solid–liquid reactions

The basic problem associated with solid–liquid reactions is one of mixing. If reactions occur at the surface of the solid there needs to be efficient transport of the reactants to the surface, and once the reaction is completed the products have to be removed. For nucleophilic substitution there is a great advantage in using a mix of an organic solvent and the salt which contains the appropriate anion.[109] Under these conditions the anion is poorly solvated and the 'naked' anion is therefore a very powerful nucleophile. This idea was exploited in the *N*-alkylation of a variety of amines, where it was found that ultrasound in conjunction with a phase-transfer catalyst accelerated the reaction and gave high yields of products.[116] Some of the results are shown in Table 3.2. More recently the method has been found of value in the *N*-methylation of diazacorands.[117]

Table 3.2 Yield of N-alkylated products from the reaction of amines with alkyl halides in toluene solution in the presence of potassium hydroxide and phase-transfer catalyst

Compound	Alkylating agent	Catalyst	Conditions	Reaction time (h)	Yield of N-alkylated product (%)
Indole	Iodomethane	PEG methyl ether	Stirred at 20°C	5	60[a,b]
Indole	Iodomethane	PEG methyl ether	Ultrasound	0·5	65[a,b]
Indole	Benzyl bromide	PEG methyl ether	Stirred at 20°C	8	80[a,b]
Indole	Benzyl bromide	PEG methyl ether	Ultrasound	2	95[a,b]
Indole	1-Bromododecane	PEG methyl ether	Stirred at room temperature	72	48[b,c]
Indole	1-Bromododecane	PEG methyl ether	Ultrasound	1·3	60[a,b]
Indole	1-Bromododecane	Tetrahexylammonium chloride	Ultrasound	1	92[a,b]
Indole	1-Bromododecane	Tetrabutylammonium nitrate	Stirred at 20°C	3	19[a,b]
Indole	1-Bromododecane	Tetrabutylammonium nitrate	Ultrasound	1·3	90[a,b]
Indole	1-Bromododecane	Tetrabutylammonium iodide	Ultrasound	0·8	73[a,b]

[a] Yield determined by GLC.
[b] No reaction observed in the absence of catalyst.
[c] Yield determined by NMR.

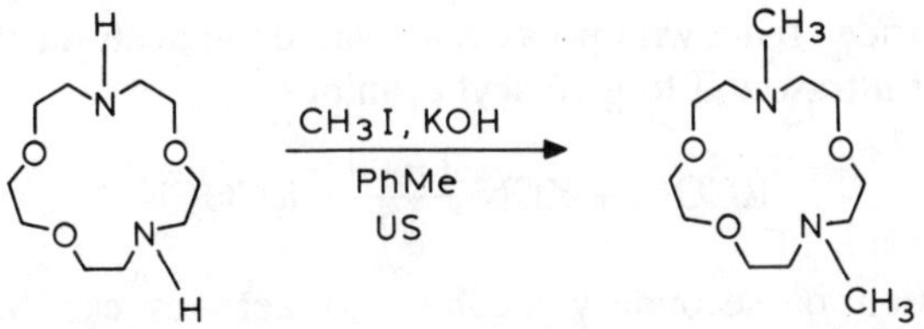

Use of the organic solvent/salt mix in conjunction with ultrasound has been applied to the preparation of esters and ethers.[109,118] The *O*-alkylation of highly hindered hydroxychromones has been accomplished using potassium carbonate as the base and *N*-methylpyrrolidone as solvent.[118] One of the functions of the ultrasound in these reactions is to break down the particle size of the potassium carbonate. With the ultrasonic probe system used in the alkylation reactions, it was found that the particle size of the potassium carbonate was reduced from $300\,\mu$m to $3\,\mu$m and that this smaller particle size material showed enhanced reactivity. Another useful heterogenous reaction is the alkylation of cyclic carbonyl compounds by α,ω-dibromoalkanes in the presence of potassium *t*-butoxide.[119] Another reaction of great synthetic value is the formation of tetrahydropyranyl-protected alcohols under mildly acidic conditions.[120] This reaction has been extended so that it can be employed to construct glycosidic bonds.

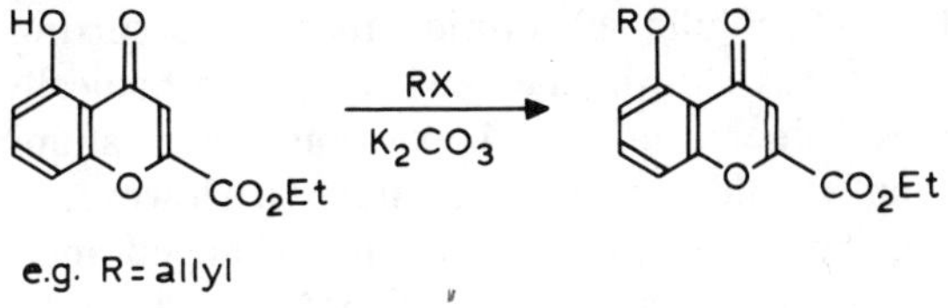

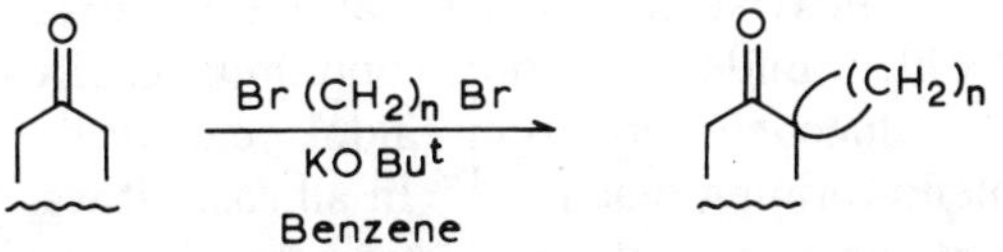

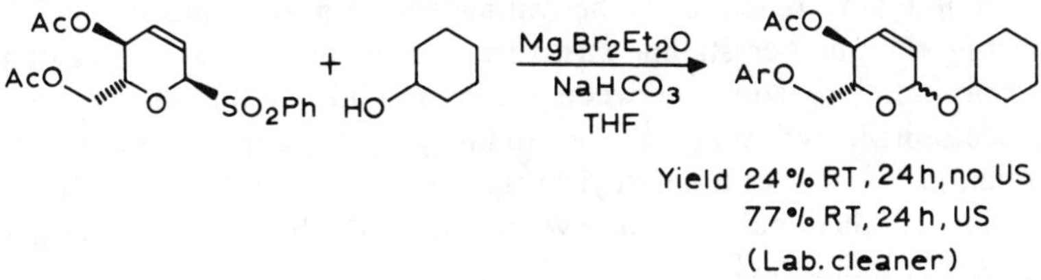

e.g. R = allyl

Yield 24 % RT, 24 h, no US
77 % RT, 24 h, US
(Lab. cleaner)

Acyl chlorides react with potassium cyanide in acetonitrile under the influence of ultrasound to give acyl cyanides.[121]

$$RCOCl + KCN \xrightarrow[US]{CH_3CN} RCOCN$$

The oxidation of secondary alcohols to ketones can be efficiently carried out using a mix of solid potassium permanganate in benzene[122] with ultrasonic agitation. Reductions with lithium aluminium hydride usually require one to work with suspensions. The reduction of aryl halides to give hydrocarbons occurs with increased efficiency when ultrasound is applied.[123] Preferential cleavage of the aryl–bromine bond over the aryl–chlorine bond is observed. Generation of silicon hydrides by reduction of alkoxysilanes has also been reported.[124]

Benzylic halides react with potassium cyanide in toluene and in the presence of alumina to give phenylacetonitriles in good yields[125] provided ultrasound is applied. Reaction also occurs with primary alkyl bromides. It was proposed that the ultrasound cleans the surface of the alumina and that catalytic sites on the alumina are responsible for the nucleophilic substitution reaction occurring. In some related work[126] it was shown that on stirring mechanically, benzyl bromide alkylates toluene in the 4 position when alumina together with potassium cyanide are used as catalysts. However, when ultrasound is applied, the course of the reaction is changed and the Friedel–Crafts reaction is suppressed and nucleophilic substitution occurs in high yield to give phenylacetonitrile. A barium hydroxide catalyst has been prepared which, under ultrasonic conditions, catalyses the Cannizzaro reaction,[127] chalcone formation (aldol reaction)[128] and the Wittig–Horner olefin-forming reaction.[129] In all cases it was proposed that the reactions occur via a single electron-transfer process and that the ultrasound aids reaction by activating sites on the catalyst and by affecting the microcrystalline structure in a beneficial way. Ultrasound also promotes reactions between the superoxide anion and chalcones in acetonitrile solution.[130] The main process appears to involve formation of the cyanomethylcarbanion, which undergoes Michael addition to the chalcone to give an adduct which then undergoes superoxide-induced cleavage.

$$O_2^{-\bullet} + CH_3CN \rightleftharpoons HO_2^{\bullet} + \bar{C}H_2CN$$

$$\bar{C}H_2CN + Ar\,CO\,CH{=}CH\,Ar \longrightarrow Ar\,\underset{\substack{\| \\ O}}{C}\,CH_2{-}\underset{\substack{| \\ CH_2CN}}{CH}\,Ar$$

$$\Big\downarrow O_2^{-}$$

$$Ar\,CO\,\bar{C}H_2 + Ar\,CH{=}CHCN$$

The formation of the observed products may also involve radical processes. The formation of thioamides by reaction of carboxamides with P_4S_{10} is another example where ultrasound aids a heterogenous reaction.[131]

A number of years ago it was reported that sonolysis of aqueous silver nitrate solutions containing added aromatic halides, e.g. bromobenzene, 2-bromothiophene, etc. yielded a precipitate which was likely to explode upon drying.[132] Although the nature of the organic products was not established, it was found that the pyrotechnic properties of the precipitate, which was in the main the appropriate silver halide, were due to the presence of up to 10% by weight of silver acetylide and diacetylide. This strongly suggests that the aromatic is suffering pyrolysis in the cavitation bubble.

Attention has been drawn to the advantages offered by ultrasound for stirring reactions more efficiently than can be attained with the normal mechanical devices. Of particular interest is the finding that the ultrasound (7·6 MHz, 1·5 kW/m^2) speeds up the reactions of the enzymes α-amylase and glucoamylase when immobilized on porous polystyrene.[133] The acceleration was attributed to the ultrasound reducing the unstirred diffusion layer around the carrier bodies. In the light of the considerable success of using polymer-supported reagents (including enzymes) this finding is of particular significance. The encroachment of the use of enzymes into classical synthetic methodology has been particularly marked in the last few years. More recently, emphasis has been placed upon the use of crude enzyme extracts. It has been shown[134] that in the case of baker's yeast the application of ultrasound (20 kHz) releases the enzyme cyclase, which transforms squalene epoxide into lanosterol, far more efficiently than conventional agitation methods and as a consequence high yields of lanosterol are realized.

Ultrasound has found use (as a result of its agitating effect) in aiding the dissolution of 3,3′,5,5′-tetramethylbenzidine in 0·1 M sulphuric acid,[135] aiding the intercalation of 1,2-dicarbadodecaborane(12) in γ-cyclodextran[136] and aiding cation exchange on the surface of charged latexes.[137]

3.4 Syntheses using metal complexes generated via ultrasound

The investigation of the reaction of metal carbonyl compounds under the influence of ultrasound has proved to be particularly rewarding, since it has not only helped to describe and define the sonochemical hot spot,[138] but has led to the generation of new chemical species.

An early discovery was that, on sonolysis, iron pentacarbonyl did not give typical thermal or photochemical products. The ultrasound induces ligand dissociation. Although the production of such species as $Fe_3(CO)_{12}$ is well established, in the sonolysis reaction these are probably produced via the intermediacy of other reactive species. Indeed, sonolysis of iron pentacarbonyl in the presence of pent-1-ene gives *cis*- and *trans*-pent-2-ene[139] via a catalytic process. Such reactions are thought to involve the alkene ligating to the metal atom and there is spectral evidence to support this proposal.[140] The ruthenium carbonyl compound $Ru_3(CO)_{12}$ is a very active catalyst and promotes the isomerization of alkenes. The product distribution in these reactions is different from that obtained from the light- and thermally catalysed reactions.[141] The metal carbonyl compounds which are particularly prone to sonolysis are those which have a sufficiently high vapour pressure to enable their vapours to be contained in the cavitational bubble without causing a significant decrease in the temperature produced on collapse of the bubbles. If halocarbons, e.g. carbon tetrachloride or chloroform, are used as solvents, homolysis of the solvent can occur, which leads to radicals capable of attacking the metal carbonyl compound.[142]

$$R_3C - Hal \rightarrow R_3C^{\cdot} + Hal^{\cdot}$$
$$2\,Hal^{\cdot} \rightarrow Hal_2$$
$$2\,Hal^{\cdot} + M_2(CO)_{10} \rightarrow 2M(CO)_5Hal$$

A most unusual complex has been reported to be formed when the metal carbonyl $Fe_2(CO)_9$ is sonolysed in the presence of excess anthracene[143] contained in *n*-hexane at $-20°$. Such a product is not formed under thermolytic or photolytic conditions.

The ability of ultrasound to remove carbon monoxide from metal carbonyl compounds has been put to good use in the formation of π-allyl carbonyl complexes. These reactions have been exploited in organic synthesis.[144,145]

Reaction of conjugated dienes in the presence of $Fe_2(CO)_9$ also gives complexes of great synthetic utility.[146]

3.5 Generation of radicals and excited states using ultrasound

The chemical effects of applying ultrasound to water have been studied extensively. It is now recognized that the process of cavitation in water can lead to its homolysis.[147-151]

$$H_2O \rightarrow HO^\cdot + H^\cdot$$

That these radicals are produced has been proved by ESR spectroscopy employing spin traps.[152]

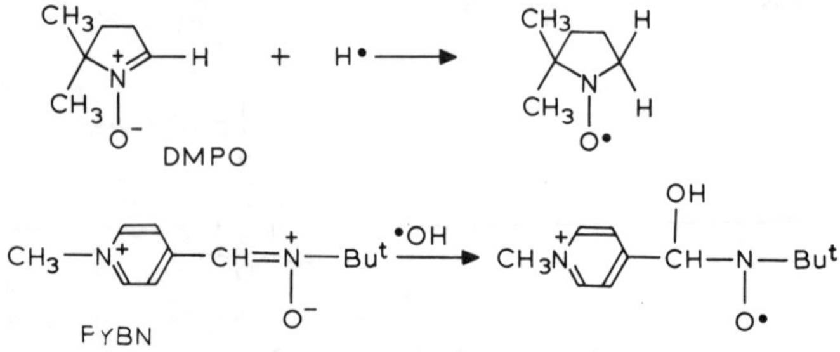

This technique was also used to show that homolysis occurs under conditions which simulate ultrasonic diagnostic conditions.[153] That the hydroxyl radicals and hydrogen atoms are primary products was also shown by spin-trapping experiments, since the addition of an appropriate radical scavenger led to a related decrease in the intensity of the signal of the spin-trapped adduct.

The formation of such reactive radicals can often give products due to secondary reactions. Thus it is recognized that hydroxyl radicals can undergo radical–radical combination reactions to give hydrogen peroxide:

$$2HO^\cdot \rightarrow H_2O_2$$

In the presence of oxygen, further reactions can take place:[154]

$$H^\cdot + O_2 \rightarrow HO_2^\cdot$$
$$2HO_2^\cdot \rightarrow H_2O_2 + O_2$$
$$HO^\cdot + H_2O_2 \rightarrow H_2O + HO_2^\cdot$$
$$HO_2^\cdot + H_2O_2 \rightarrow O_2 + HO^\cdot + H_2O$$

The amount of hydrogen peroxide produced upon sonolysing water as a function of hydrogen and oxygen concentration has been measured.[154] The combustion reaction between hydrogen and oxygen occurs in the cavitational bubble with the attendant chain processes. However, once the radicals have reached the bubble–water interface, the chain processes cease. The distinction between what happens within the bubble and at the interface and in solution cannot be too strongly emphasized. Use of various hydrogen-atom scavengers, e.g. DMPO, iodine, permanganate anions, led to the desired trapping, but in the case of DMPO the yield of trapped radicals was found to be very low.[148] This led to the important suggestion that the efficiency of the trapping process was related to the ability of the trap to accumulate at the water–bubble interface. There have been reports that sonolysis of water generates hydrated electrons. That hydrated electrons are produced has been verified;[148] however, they are not produced in a primary process but rather as a result of a secondary reaction:

$$H^{\cdot} + \bar{O}H \rightarrow H_2O + e_{aq}$$

Sonolysis of aqueous solutions of ozone accelerates the thermal decomposition of the latter.[155] Decomposition is facilitated by ozone's presence in the bubble together with such gases as argon. The function of the latter is to increase the temperature attained on collapse of the bubble. Decomposition of ozone produced oxygen atoms and reactions of these species were identified:

$$O_3 \rightarrow O_2 + O$$
$$O + H_2O \rightarrow 2\dot{O}H$$
$$H^{\cdot} + O_2 \rightarrow HO_2^{\cdot}$$
$$2HO_2^{\cdot} \rightarrow H_2O_2 + O_2$$

Sonolysis of water containing a mixture of deuterium and methane (2:1 by volume) led to isotopic exchange.[156] In addition a range of carbon-containing products was formed, e.g. ethane, ethene, ethyne and higher hydrocarbons, suggesting that some pyrolysis of the methane occurred. The exchange reactions are thought to be due to reaction of methane with hydroxyl radicals etc.:

$$H_2O \rightarrow HO^{\cdot} + H^{\cdot}$$
$$HO^{\cdot} + D_2 \rightarrow HDO + D^{\cdot}$$
$$HO^{\cdot} + CH_4 \rightarrow \dot{C}H_3 + H_2O$$
$$H^{\cdot} + D_2 \rightarrow HD + D^{\cdot}$$
$$CH_4 \rightarrow \dot{C}H_3 + H^{\cdot}$$
$$\dot{C}H_3 + D_2 \rightarrow CH_3D + D^{\cdot}$$

Sonolysis of water containing deuterated formate ions produced HD, which is further evidence for the intermediacy of hydrogen atoms:[157]

$$H_2O \rightarrow H^{\cdot} + \dot{O}H$$
$$H^{\cdot} + DCO_2^- \rightarrow HD + CO_2^{\overset{\cdot}{-}}$$

Introduction of aromatic hydrocarbons (e.g. toluene) into the aqueous solution leads to hydroxylation—indicative that hydroxyl radicals are generated upon sonolysis.[158]

Sonolysis of aqueous solutions of di-n-butyl sulphide at a variety of frequencies produced the sulphoxide as the main product.[159] In addition a little sulphone was produced and both the sulphoxide and sulphone gave, upon extensive sonolysis in the presence of oxygen, n-butyl sulphenic acid and n-butyraldehyde. It was suggested that the formation of the sulphoxide occurred via electron ejection from the sulphide. Though this is not disproved there is the clear possibility that hydrogen peroxide generated from the water is responsible for oxidation. Sonolysis of the aqueous oxygenated solutions of the sulphide yielded unidentified polymeric materials.

Very strong evidence for the formation of hydroxyl radicals upon sonolysis of water comes from a detailed study of the products formed upon sonolysis of aqueous solutions of nucleic acid bases. The reactivity of the bases was found to be thymine > uracil > cytosine > guanine > adenine.[160] Not surprisingly, it was found that the reactions were most efficient when the solutions were saturated with argon, with lower efficiencies being observed when diatomic gases were present. The main products from thymine have been shown to be 5-hydroxymethyluracil and 5,6-dihydroxy-5,6-dihydrothymine.[161] Formation of the latter indicates that addition of hydroxyl radicals to the heterocycle is an important reaction.[162]

Mixture of isomers

Similar processes occur with uracil, but in this case the studies revealed the formation of ring-cleavage and ring-contraction products.[163]

The extent to which hydrogen peroxide mediates in product formation is not clear—e.g. is the 5,6-epoxide of the pyrimidine an important intermediate? Sonolysis of aqueous solutions of aliphatic aldehydes under argon leads to the production of the appropriate carboxylic acid.[64] If the reaction is carried out in deuterium oxide, deuterium exchange on the α-carbon atom occurs. The reactions were interpreted as involving electron ejection from the aldehyde in the collapsing cavitation bubble. However, it is possible to rationalize product formation on the basis of the aldehyde reacting with the sonolysis products of water:

$$RCHO \xrightarrow{H^{\cdot} \text{ or } OH} R\dot{C}O$$

$$R\dot{C}O + H_2O_2 \rightarrow RCO_2H + H\dot{O}$$

$$R\dot{C}O \rightarrow R^{\cdot} + CO$$

$$R^{\cdot} + RCHO \rightarrow RH + R\dot{C}O$$

$$-CH_2CHO \xrightarrow{H^{\cdot} \text{ or } OH} -\dot{C}HCHO$$

$$-CH_2CO_2H \xrightarrow{H^{\cdot} \text{ or } OH} -\dot{C}HCO_2H$$

The fact that sonolysis of water generates hydrogen peroxide means that, if aqueous solutions of iron (II) salts are sonolysed, reactions typical of Fentons' reagent occur. This has been observed for the oxidation of secondary alcohols upon sonolysis of their aqueous solutions containing iron (II) salts.[165]

The formation of hydrogen peroxide can readily be detected by its reaction with potassium iodide to liberate iodine.[166] Addition of carbon tetrachloride led to an increase in the yield of iodine. This was attributed to the sonolysis of carbon tetrachloride giving chlorine.[142] It was also found that sonolysis of dilute hydrochloric acid (a decomposition product of carbon tetrachloride) also produced chlorine. It is well known that hydrogen peroxide reacts with many α-amino acids (cysteine, cystine, methionine, tryptophan, etc.) and therefore it is not surprising to find that sonolysis of aqueous solutions of α-amino acids leads to their destruction.[167] Not surprisingly, cystine is oxidized to cysteic acid very efficiently and cysteine is converted to cystine. Other amino acids are less prone to degradation but nevertheless do undergo decomposition. Another particularly sensitive way of detecting hydrogen peroxide is by its chemiluminescent reaction with luminol.

Very early on it was established that oxygen had to be present in the reaction mixture for chemiluminescence to be observed.[168,169] The ultrasonically aided reaction has been incorporated into an analytical method for detection of sub μg levels of cobalt (II) ions.[170]

Other sonolysis reactions of organic materials, e.g. allylamine[171] and D-glucose[172] have been described, but a mechanistic understanding of what is occurring is lacking.

Work on the sonolysis of organic materials did not progress for many years owing to the fact that complex reaction mixtures were obtained (e.g. acetonitrile gives nitrogen, methane and hydrogen[173]). Many aromatic compounds undergo discoloration upon sonolysis and this has been interpreted as the aromatics undergoing polymerization.[174] Whatever the nature of the products, it seems clear that they are generated by something akin to a pyrolysis process which must occur on collapse of the cavitational bubble.[175] Needless to say, carbon tetrachloride homolyses upon sonolysis[142,175] and spectroscopic evidence for radical formation[176] has been obtained. A very detailed study has been made of the decomposition of chloroform.[177] Homolysis occurs and the reactions identified were as follows:

$$CHCl_3 \rightarrow \dot{C}HCl_2 + \dot{C}l$$
$$\dot{C}Cl_3 + H^{\cdot}$$
$$:CCl_2 + HCl$$
$$:CHCl + Cl_2$$

Decreasing probability of occurrence

Products formed via these processes included carbon tetrachloride, tetrachloroethane, penta- and hexachloroethane and tetrachloroethene. Unambiguous evidence for carbene formation came from studies in which a chloroform solution of cyclohexene was sonolysed. The olefin is sufficiently volatile to be present in the cavitational bubble and traps the carbene with formation of the cyclopropane derivatives. Sonication of styrene leads to a variety of products.[178] Formation of highly coloured products is favoured by conditions which lead to something akin to pyrolysis of styrene, whereas, if the temperature produced during cavitational bubble collapse is reduced by adding material of fairly high vapour pressure, the yield of polystyrene is increased at the expense of the unidentified coloured materials. That many materials undergo thermal degradation is not surprising after one considers the temperatures attainable by bubble collapse (>5000 K).[138] Further evidence that thermal decomposition occurs on bubble collapse comes from sonoluminescence studies[179] in which it was shown that the emission produced from hydrocarbons (e.g. dodecane) is due to the $d^3\pi g \rightarrow a^3\pi u$ transition for C_2. Clearly, processes similar to thermal cracking and those which occur under shock-tube conditions are attained during sonolysis.

Mixtures of organic compounds have been subjected to ultrasound. A particularly useful application has been the ultrasound-assisted addition of methyl disulphide to hexafluorobutadiene,[180] which involves homolysis of the disulphide. Sonolysis of acetonitrile solutions of aromatic diazonium salts leads to the homolytic decomposition of the latter and the radicals produced have been identified using spin-trapping techniques.[181]

When solutions of metal carbonyl compounds in halocarbons are subjected to ultrasound, the primary reaction appears to be homolysis of the solvent followed by radical attack upon the metal–carbonyl bond (e.g. $Mn_2(CO)_{10}$ and $Re_2(CO)_{10}$).[142]

The presence of oxygen in organic solvents subjected to ultrasound raises the possibility of oxidation products being formed. This has been realized in the case of alkenes which generate allylic hydroperoxides, which in the presence of molybdenum hexacarbonyl give allylic alcohols and oxiranes.[182]

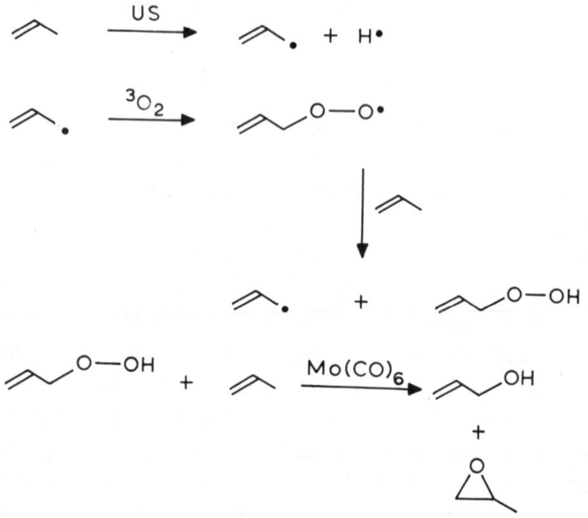

From the observation of sonoluminescence it can be concluded that excited states are present within the collapsing cavitational bubble. If

Table 3.3 Kinetic solvent isotope effects for ultrasound-initiated oxidation of anthracenes[a] and indoles[a]

Substrate	Solvent system	
	$CDCl_3/CHCl_3$	CD_3OD/CH_3OH
Anthracene	3·8	2·8
9-Methylanthracene	4·5	3·6
9,10-Dimethylanthracene	6·8	4·5
Indole	5·6	a
3-Methylindole	7·2	b
2,3-Dimethylindole	8·5	b

[a] Concentration 1×10^{-4} M.
[b] Not determined.

oxygen is present in the bubble, there is the possibility that an excited state of oxygen may be formed (e.g. $^1\Delta_g O_2$). Indeed, sonication of chloroform solutions of anthracenes leads to the formation of anthra-9,10-quinone.[109] If deuterochloroform is used in place of chloroform, there is a rate enhancement. Such a kinetic solvent isotope effect is indicative of the participation of singlet oxygen. Other materials, such as indoles, are oxidized under similar conditions and display kinetic solvent isotope effects (Table 3.3).

From the foregoing it would appear that many more ultrasound-initiated free-radical reactions (1,2-addition and polymerization reactions) are awaiting exploitation.

3.6 References

1. W. Slough & A. R. Ubbelhode, *J. Chem. Soc.*, 918 (1957).
2. M. W. T. Pratt & R. Helsby, *Nature (London)*, **184**, 1694 (1959).
3. T. Azuma, S. Yanagida, H. Sakurai, S. Sasa & Y. Yoshino, *Synth. Commun.* **12**, 137 (1982).
4. A. J. Birch, *J. Chem. Soc.*, 1642 (1947); see also A. J. Birch & H. Smith, *Quart. Rev.*, **XII**, 17 (1958).
5. P. Boudjouk, R. Sooriyakumaran & B. H. Han, *J. Org. Chem.*, **51**, 2818 (1986).
6. J. W. Huffman, W.-P. Liao & R. H. Wallace, *Tetrahedron Lett.*, **28**, 3315 (1987).
7. J. Einhorn, C. Einhorn & J.-L. Luche, *Tetrahedron Lett.* **29**, 2183 (1988).

108　　　*R. S. Davidson*

8. J.-L. Luche & J. C. Damiano, *J. Am. Chem. Soc.*, **102**, 7926 (1980).
9. J. Einhorn & J.-L. Luche, *J. Org. Chem.*, **52**, 4124 (1987).
10. J. C. de Souza-Barboza, J.-L. Luche & C. Pétrier, *Tetrahedron Lett.*, **28**, 2013 (1987).
11. T. D. Lash & D. Berry, *J. Chem. Ed.*, **62**, 85 (1985).
12. B. M. Trost & B. P. Coppola, *J. Am. Chem. Soc.*, **104**, 6879 (1982).
13. C. Petrier, A. L. Gemal & J.-L. Luche, *Tetrahedron Lett.*, **23**, 3361 (1982).
14. J. Einhorn & J.-L. Luche, *Tetrahedron Lett.*, **27**, 1791 (1986).
15. J. Einhorn & J.-L. Luche, *Tetrahedron Lett.*, **27**, 1793 (1986).
16. J. Einhorn, J.-L. Luche & P. Demerseman, *J. C. S. Chem. Commun.*, 1350 (1988).
17. B. H. Han & P. Boudjouk, *Tetrahedron Lett.*, **22**, 2757 (1981).
18. J. Einhorn & J.-L. Luche, *Tetrahedron Lett.*, **27**, 501 (1986).
19. J. Einhorn & J.-L. Luche, *J. Org. Chem.*, **52**, 4124 (1987).
20. R. West, A. R. Wolff & D. J. Peterson, *J. Radiation Curing*, **13**, 35 (1986); idem. Polysilanes: A New Class of Vinyl Photoinitiators, *18th Organosilicon Symposium*, Schenectedy, NY, 14 April, 1984.
21. J. Michl, J. W. Downing, T. Karatsu, A. J. McKinley, G. Poggi, G. M. Wallraff, R. Sooriyakumaran & D. R. Miller, *Pure Appl. Chem.*, **60**, 959 (1988).
22. P. Boudjouk, *J. Chem. Ed.*, **63**, 427 (1986).
23. P. Boudjouk & B. H. Han, *Tetrahedron Lett.*, **22**, 3813 (1981).
24. P. Boudjouk, B. H. Han & K. R. Anderson, *J. Am. Chem. Soc.*, **104**, 4992 (1982).
25. S. Masamune, S. Murakami & H. Tobita, *Organometallics*, **2**, 1464 (1983).
26. C. Eaborn, P. B. Hitchcock & P. D. Lickiss, *J. Organometall. Chem.*, **269**, 235 (1984).
27. T.-S. Chou, J.-J. Yuan & C.-H. Tsao, *J. Chem. Res. (S)*, 18 (1985).
28. T.-S. Chou, C.-H. Tsao & S. C. Hung, *J. Org. Chem.*, **50**, 4329 (1985).
29. J.-L. Luche, C. Petrier & C. Dupuy, *Tetrahedron Lett.*, **25**, 753 (1984).
30. S. V. Ley, I. A. O'Neil & C. M. R. Low, *Tetrahedron*, **42**, 5363 (1986).
31. O. de Lucchi, N. Piccolrovazzi & G. Modena, *Tetrahedron Lett.*, **27**, 4347 (1986).
32. T.-S. Chou & M.-L. You, *Tetrahedron Lett.*, 4495 (1985).
33. T.-S. Chou & M.-L. You, *J. Org. Chem.*, **52**, 2224 (1987).
34. P. A. Bianconi & T. W. Weidman, *J. Am. Chem. Soc.*, **110**, 2342 (1988).
35. K. Sjoberg, *Tetrahedron Lett.*, 6383 (1966).
36. P. Renaud, *Bull. Soc. Chim. Fr. Ser. S*, **17**, 1044 (1950).
37. J. P. Sprich & G. Lewandos, *Inorg. Chim. Acta.*, **76**, L241 (1983).
38. W. Oppolzer & P. Schneider, *Tetrahedron Lett.*, **25**, 3305 (1984).
39. R. Dharanipragada & F. Diederich, *Tetrahedron Lett.*, **28**, 2443 (1987).
40. M. Bonneman, B. Bogdanovic, R. Brinkman, D. W. He & B. Spliethoff, *Angew. Chem. Int. Ed.*, **22**, 728 (1983).
41. G. Casiraghi, M. Cornia, G. Casnati, G. G. Fava, M. F. Belicchi & L. Zetta, *J. C. S. Chem. Commun.*, 794 (1987).
42. N. Ishikawa, M. G. Koh, T. Kitazume & S. K. Choi, *J. Fluorine Chem.*, **24**, 419 (1984).

43. G. W. Kabalka, R. S. Varma, Y.-Z. Gai & R. M. Baldwin, *Tetrahedron Lett.*, **27**, 3843 (1986).
44. H. C. Brown & U. S. Racherla, *Tetrahedron Lett.*, **26**, 4311 (1985).
45. T. Kitazume & N. Ishikawa, *Chem. Lett.*, 1679 (1981).
46. T. Kitazume & N. Ishikawa, *J. Am. Chem. Soc.*, **107**, 5186 (1985).
47. M. Fujita, T. Morita & T. Hiyama, *Tetrahedron Lett.*, **27**, 2135 (1986).
48. A. Solladie-Cavello, D. Farkhani, S. Fritz, T. Lazrak & J. Suffert, *Tetrahedron Lett.*, **25**, 4117 (1984).
49. B. H. Han & P. Boudjouk, *J. Org. Chem.*, **47**, 5030 (1982).
50. R. W. Lang & B. Schaub, *Tetrahedron Lett.*, **29**, 2943 (1988).
51. N. Ishikawa, M. G. Koh, T. Kitazumi & S. K. Choi, *J. Fluorine Chem.*, **24**, 419 (1984).
52. T. Kitazume, *Synthesis*, 855 (1986).
53. A. K. Bose, K. Gupta & M. S. Manhas, *J. C. S. Chem. Commun.*, 86 (1984).
54. J. Brennan & F. H. S. Hussain, *Synthesis*, 749 (1985).
55. P. Knochel & J. F. Normant, *Tetrahedron Lett.*, **25**, 1475 (1984).
56. C. Pétrier & J.-L. Luche, *J. Org. Chem.*, **50**, 910 (1985).
57. J.-L. Luche, C. Pétrier, J. P. Lansard & A. E. Greene, *J. Org. Chem.*, **48**, 3837 (1983).
58. C. Pétrier, J.-L. Luche & C. Dupuy, *Tetrahedron Lett.*, **25**, 3463 (1984).
59. A. E. Greene, J. P. Lansard, J.-L. Luche & C. Pétrier, *J. Org. Chem.*, **49**, 931 (1984).
60. J. C. de Souza Barboza, C. Petrier & J.-L. Luche, *Tetrahedron Lett.*, **26**, 829 (1985).
61. J.-L. Luche & C. Allavena, *Tetrahedron Lett.*, **29**, 5369 (1988).
62. K. M. Pietrusiewicz & M. Zablocka, *Tetrahedron Lett.*, **29**, 937 (1988).
63. C. Pétrier, C. Dupuy & J.-L. Luche, *Tetrahedron Lett.*, **27**, 3149 (1986).
64. J.-L. Luche, C. Allavena, C. Petrier & C. Dupuy, *Tetrahedron Lett.*, **29**, 5373 (1988).
65. H.-H. Tso, T.-S. Chou & S. C. Hung, *J. C. S. Chem. Commun.*, 1552 (1987).
66. C. Pétrier & J.-L. Luche, *Tetrahedron Lett.*, **28**, 2347, 2351 (1987).
67. P. Boudjouk & J. Ho So, *Synth. Commun.* **16**, 775 (1986).
68. G. Mehta & H. S. P. Rao, *Synth. Commun.*, **15**, 991 (1985).
69. N. N. Joshi and H. M. R. Hoffmann, *Tetrahedron Lett.*, **27**, 687 (1986).
70. A. J. Fry, G. S. Ginsburg, & R. A. Parante, *J. C. S. Chem. Commun.*, 1040 (1978).
71. A. J. Fry & G. S. Ginsburg, *J. Amer. Chem. Soc.*, **101**, 3927 (1979).
72. H. C. Brown & U. S. Racherla, *Tetrahedron Lett.*, **26**, 2187 (1985).
73. Y.-T. Lin, *J. Organometall. Chem.*, **317**, 277 (1986).
74. K. F. Liou, P.-H. Yang & Y.-T. Lin, *J. Organometall. Chem.*, **294**, 145 (1985).
75. B. M. Trost & B. P. Coppola, *J. Am. Chem. Soc.*, **104**, 6879 (1982).
76. P. Boudjouk, W. H. Ohrbram & J. B. Woell, *Synth. Commun.*, **16**, 401 (1986).
77. T. Ando, S. Sumi, T. Kawate, J. Ichihara & T. Hanafusa, *J. C. S. Chem. Commun.*, 439 (1984).
78. G. Saraceo & A. Arzano, *Chim. Ind.* (*Milano*), **50**, 314 (1968).

79. K. J. Moulton, S. Koritala & E. N. Frankel, *J. Am. Oil Chem. Soc.*, **60**, 1257 (1983).
80. K. J. Moulton, S. Koritala, K. Warner & E. N. Frankel, *J. Am. Oil Chem. Soc*; **64**, 542 (1987).
81. E. A. Cioffi & J. H. Prestegard, *Tetrahedron Lett.*, **27**, 415 (1986).
82. P. Boudjouk & H.-B. Han, *J. Catal.*, **79**, 489 (1983).
83. B.-H. Han & P. Boudjouk, *Organometallics*, **2**, 769 (1983).
84. K. Imi, K. Imai & K. Utimoto, *Tetrahedron Lett.*, **28**, 3127 (1987).
85. J. Lindley, J. P. Lorimer & T. J. Mason, *Ultrasonics*, **24**, 292 (1986).
86. J. Yamashita, Y. Inoue, T. Kondo & H. Hashimoto, *Chem. Lett.*, 407 (1986).
87. O. Repic & S. Vogt, *Tetrahedron Lett.* **23**, 2729 (1982).
88. L. Xu, F. Tao & T. Yu, *Tetrahedron Lett.*, **26**, 4231 (1985).
89. O. Repic, P. G. Lee & N. Giger, *Org. Prep. Proc. Int.*, **16**, 25 (1984).
90. S. L. Regen & A. Singh, *J. Org. Chem.*, **47**, 1587 (1982).
91. M. S. F. L. K. Jie & W. L. K. Lam, *J. C. S. Chem. Commun.*, 1460 (1987).
92. C. Einhorn, C. Allavena & J.-L. Luche, *J. C. S. Chem. Commun.*, 333 (1988).
93. B. H. Han & P. Boudjouk, *J. Org. Chem.*, **47**, 751 (1982).
94. C. W. Porter & L. Young, *J. Am. Chem. Soc.*, **60**, 1497 (1938).
95. T. J. Mason & J. P. Lorimer, *J. C. S. Chem. Commun.*, 1135 (1980).
96. T. J. Mason, J. P. Lorimer & B. P. Mistry, *Tetrahedron Lett.*, **23**, 5363 (1982).
97. T. J. Mason, J. P. Lorimer & B. P. Mistry, *Tetrahedron Lett.*, **24**, 4371 (1983).
98. T. J. Mason, J. P. Lorimer & B. P. Mistry, *Tetrahedron*, **41**, 5201 (1985).
99. C. Einhorn & J.-L. Luche, *Carbohydrate Res.*, **155**, 258 (1986).
100. S. J. Hatin, S. H. Kim, H. J. Chae, B. H. Young & H. S. Lyu, *Bull. Korean Chem. Soc.*, **8**, 49 (1987).
101. S. Toma, M. Putala & M. Salisoua, *Coll. Czech. Chem. Commun.*, **52**, 395 (1987).
102. D. R. Borthakur & J. S. Sandhu, *J. C. S. Chem. Commun.*, 1444 (1988).
103. J. C. Menendez, G. G. Trigo & M. M. Sollhuber, *Tetrahedron Lett.*, **27**, 3285 (1986).
104. U. Schuchardt, I. Joekes & H. C. Duarte, *J. Chem. Tech. Biotechnol.*, **39**, 115 (1987).
105. Y. Yoshino, S. Paoletti, J. Kido & Y. Okamato, *Macromolecules*, **18**, 1513 (1985).
106. G. P. Spada, P. Brigidi & G. Gottarelli, *J. C. S. Chem. Commun.*, 953 (1988).
107. D. S. Kristol, H. Klotz & R. C. Parker, *Tetrahedron Lett.*, **22**, 907 (1981); I. Miyagawa, *J. Soc. Org. Synth. Commun.* (Japan), **7**, 167 (1949).
108. S. Moon, L. Duchin & J. V. Cooney, *Tetrahedron Lett.*, 3917 (1979).
109. R. S. Davidson, A. Safdar, J. D. Spencer & B. Robinson, *Ultrasonics*, **25**, 35 (1987).
110. R. S. Davidson, A. Safdar & B. Robinson, UK Patent Application, 8602316, 1986.

111. J. Ezquerra & J. Alvarez-Builla, *Org. Prep. Proc. Int.*, **17**, 190 (1985).
112. R. Neumann & Y. Sasson, *J. C. S. Chem. Commun.*, 616 (1985).
113. J. E. Ezquerra & J. Alvarez-Builla, *J. C. S. Chem. Commun.*, 54 (1984).
114. W. P. Reeves & J. V. McClusky, *Tetrahedron Lett.*, 1585 (1983).
115. H. Priebe, *Acta Chem. Scand., Ser. B*, **38**, 895 (1984).
116. R. S. Davidson, A. M. Patel, A. F. Safdar & D. Thornthwaite, *Tetrahedron Lett.*, **24**, 5907 (1983).
117. J. Jurczak & R. Ostaszewski, *Tetrahedron Lett.*, **29**, 959 (1988).
118. J. P. Lorimer, T. J. Mason, A. T. Turner & A. R. Harris, *J. Chem. Res. (S)*, 80 (1988).
119. T. Fujita, S. Watawabe, M. Sakamoto & H. Hashimoto, *Chem. Ind.*, 427 (1986).
120. D. S. Brown, S. V. Ley & S. Vile, *Tetrahedron Lett.*, **29**, 4873 (1988).
121. T. Ando, T. Kawate, J. Yamawaki & T. Hanafusa, *Synthesis*, 637 (1983).
122. J. Yamawaki, S. Sumi, T. Ando & T. Hanafusa, *Chem. Lett.*, 379 (1983).
123. B. H. Han & P. Boudjouk, *Tetrahedron Lett.*, **23**, 1643 (1982).
124. E. Lukevics, V. N. Gevorgyan & Y. S. Goldberg, *Tetrahedron Lett.*, **25**, 1415 (1984).
125. T. Ando, T. Kawate, J. Ichihara & T. Hanafusa, *Chem. Lett.*, 725 (1984).
126. T. Ando, S. Sumi, T. Kawate, J. Ichihara & T. Hanafusa, *J. C. S. Chem. Commun.*, 439 (1984).
127. A. Fuentes, J. M. Marinas & J. V. Sinisterra, *Tetrahedron Lett.*, **28**, 2947 (1987).
128. A. Fuentes, J. M. Marinas & J. V. Sinisterra, *Tetrahedron Lett.*, **28**, 4541 (1987).
129. J. V. Sinisterra, A. Fuentes & J. M. Marinas, *J. Org. Chem.*, **52**, 3875 (1987).
130. K. Shibata, K. Urano & M. Matsui, *Chem. Lett.*, 519 (1987).
131. S. Raucher & P. Klein, *J. Org. Chem.*, **46**, 3558 (1981).
132. L. Zeichmeister & L. Wallcave, *J. Am. Chem. Soc.*, **77**, 2853 (1955).
133. P. Schmidt, E. Rosenfield, R. Millner, R. Czerner & A. Schellen Berger, *Biotechnol. Bioeng.*, **30**, 928 (1987).
134. J. Bujons, R. Guajardo & K. S. Kyler, *J. Am. Chem. Soc.*, **110**, 604 (1988).
135. A. Watanabe, K. Mori, M. Mikuni, Y. Nakamura & O. Ito, *J. C. S. Chem. Commun.*, 452 (1988).
136. A. Harada & S. Takahashi, *J. C. S. Chem. Commun.*, 1352 (1988).
137. R. S. Chandran & W. T. Ford, *J. C. S. Chem. Commun.*, 104 (1988).
138. K. S. Suslick, D. A. Hammerton & R. E. Cline, *J. Am. Chem. Soc.*, **108**, 5641 (1986).
139. K. S. Suslick, P. F. Schubert & J. W. Goodale, *J. Am. Chem. Soc.*, **103**, 7342 (1981).
140. K. S. Suslick, J. W. Goodale, P. F. Schubert & H. H. Wang, *J. Am. Chem. Soc.*, **105**, 5781 (1983).
141. K. S. Suslick, *Modern Synthetic Methods*, **41**, 1 (1986).

142. K. Suslick & P. F. Schubert, *J. Am. Chem. Soc.*, **105**, 6042 (1983).
143. M. J. Begley, S. G. Puntambekar & A. H. Wright, *J. C. S. Chem. Commun.*, 1251 (1987).
144. A. M. Horton, D. M. Hollinshead & S. V. Ley, *Tetrahedron*, **40**, 1737 (1984).
145. A. M. Horton & S. V. Ley, *J. Organometall. Chem.*, **285**, C17 (1985).
146. S. V. Ley, C. M. R. Lowe & A. D. White, *J. Organometall. Chem.*, **302**, C13 (1986).
147. Ya I. Frenkel, *Zh. Fiz., Khim.*, **14**, 305 (1940).
148. M. Gutierrez, A. Henglein & J. K. Dohrmann, *J. Phys. Chem.* **91**, 6687 (1987).
149. A. Henglein and C. Kormann, *Int. J. Rad. Biol.* **48**, 251 (1985).
150. C. Sehgal, T. J. Yu, R. G. Sutherland & R. E. Verrall, *J. Phys. Chem.*, **86**, 2982 (1982).
151. P. Riesz, D. Berdahal & C. L. Christman, *Environ. Health Perspect.*, **64**, 233 (1985).
152. K. Makino, M. M. Mossoba & P. Riesz, *J. Am. Chem. Soc.*, **104**, 3537 (1982).
153. A. J. Carmichael, M. M. Mossoba, P. Riesz & C. L. Christman, *IEEE Trans. Ultrasonics, Ferroelectronics and Frequency Control*, **UFFC-33**, 148 (1986).
154. E. J. Hart & A. Henglein, *J. Phys. Chem.*, **91**, 3654 (1987).
155. E. J. Hart & A. Henglein, *J. Phys. Chem.*, **90**, 3061 (1986).
156. E. J. Hart, C. H. Fischer & A. Henglein, *J. Phys. Chem.*, **91**, 4166 (1987).
157. M. Anbar & I. Pecht, *J. Phys. Chem.*, **68**, 1460 (1964).
158. A. V. Parke & D. Taylor, *J. Chem. Soc.*, 4442 (1956).
159. L. A. Spurlock & S. B. Reifsnider, *J. Am. Chem. Soc.*, **92**, 6112 (1970).
160. J. R. McKee, C. L. Christman, W. D. O'Brien & S. Y. Wang, *Biochemistry*, **16**, 4651 (1977).
161. E. L. Mead, R. G. Sutherland & R. E. Verrall, *Can. J. Chem.*, **53**, 2394 (1975).
162. C. M. Sehgal & S. Y. Wang, *J. Am. Chem. Soc.*, **103**, 6606 (1981).
163. T. J. Yu, R. G. Sutherland & R. E. Verrall, *Can. J. Chem.*, **58**, 1909 (1980).
164. S. B. Reifsneider & L. A. Spurlock, *J. Am. Chem. Soc.*, **95**, 299 (1973).
165. C. Sehgal, R. G. Sutherland & R. E. Verrall, *J. Phys. Chem.*, **84**, 2920 (1980).
166. A. Weissler, H. W. Cooper & S. Snyder, *J. Am. Chem. Soc.*, **72**, 1769 (1950).
167. W. H. Staas & L. A. Spurlock, *J. C. S. Perkin 1*, 1675 (1975).
168. E. W. Flosdorf, L. A. Chambers & W. M. Malisoff, *J. Am. Chem. Soc.*, **58**, 1069 (1936).
169. E. N. Harvey, *J. Am. Chem. Soc.*, **61**, 2392 (1939).
170. T. Komatsu, M. Ohira, M. Yamada & S. Suzuki, *Bull. Chem. Soc., Japan*, **59**, 1849 (1986).
171. S. Nishikawa, U. Otani & M. Mashima, *Bull. Chem. Soc., Japan*, **50**, 1716 (1977).

172. H. Heusinger, *Carbohydrate Res.*, **154**, 37 (1986).
173. A. Weissler, I. Pecht & M. Anbar, *Science*, **150**, 1288 (1965).
174. D. J. Donaldson, M. D. Farrington & P. Kruus, *J. Phys. Chem.*, **83**, 3130 (1979).
175. P. Kruus, *Ultrasonics*, **25**, 20 (1987).
176. I. Rosenthal, M. M. Mossoba & P. Reisz, *J. Magn. Reson.*, **45**, 359 (1981).
177. A. Henglein & Ch.-H. Fischer, *Ber. Bunsenges. Phys. Chem.*, **88**, 1196 (1984).
178. P. Kruus, D. McDonald & T. J. Patraboy, *J. Phys. Chem.*, **91**, 3041 (1987).
179. K. S. Suslick & E. B. Flint, *Nature*, **330**, 553 (1987).
180. M. S. Toy & R. S. Stringham, *J. Flourine Chem.*, **29**, 253 (1985).
181. D. Rehorek & E. G. Janzen, *J. Prakt. Chem.*, **326**, 935 (1984).
182. K. S. Suslick, *High Energy Processes in Organometallic Chemistry*. ACS Symposium No. 333.

4 Polymers

J. P. Lorimer
Coventry Polytechnic, UK

Polymer chemists have at their disposal two types of ultrasonic wave: high-intensity (usually low-frequency) and low-intensity (usually high-frequency). The use of low-intensity waves provides information on relaxation phenomena such as segmental motion, conformational change, vibrational–translational energy interchange and polymer–solvent interactions, whilst high-intensity waves have been used to improve polymer processing and to effect such chemical changes as polymerization and depolymerization. In order to appreciate more fully the enormous impact of ultrasound on polymer systems it is appropriate to remind ourselves of the important equations involved with the passage of ultrasound.

4.1 Introduction

It may be recalled (Chapter 1, Section 1.3.1) that the total pressure, P_T, in a liquid at any time, as a result of the passage through it of a

116 J. P. Lorimer

sound wave, is given by

$$P_T = P_h + P_a \tag{4.1}$$

where P_h is the ambient or atmospheric pressure present in the liquid, and P_a is the acoustic pressure created by the sonic vibrations (eqn (4.2)),

$$P_a = P_A \sin 2\pi ft \tag{4.2}$$

The effect of this acoustic pressure wave, as it propagates through the medium, is to induce oscillations of the molecules about their mean rest position (eqn (4.3)) and thereby increase, momentarily, their mean translational energy (e.g. $\frac{1}{2}mv^2$, where v is given by eqn (4.4)):

$$x = x_0 \sin 2\pi ft \tag{4.3}$$

$$v = \frac{dx}{dt} = 2\pi f x_0 \cos 2\pi ft \tag{4.4}$$

Although in principle this translational energy can be transferred totally by elastic collisions to other molecules, and so increase their translational energy, it has been suggested that in reality energy losses will occur due to thermal and frictional losses. The extent of this attenuation has already been shown to be represented by the following expression:

$$I = I_0 \exp(-2\alpha l) \tag{4.5}$$

where α is the absorption (attenuation) coefficient, I_0 is the initial intensity of the sound wave and I is the intensity at some distance l from the source and related to the maximum acoustic pressure, P_A, by eqn (4.6)

$$I = \frac{P_A^2}{2\rho c} \tag{4.6}$$

According to Stokes[1] and Kirchoff,[2] the absorption coefficients due to the frictional (α_s) and the thermal (α_{th}) losses occurring on the passage of the wave are given respectively by

$$\alpha_s = \frac{8\eta\pi^2 f^2}{3\rho c^3} \tag{4.7}$$

and

$$\alpha_{th} = \frac{2\pi^2 K(\gamma - 1)f^2}{\rho\gamma C_v c^3} \tag{4.8}$$

where η is the viscosity of the medium,
  f is the frequency of the sound source,
  ρ is the density of the medium,
  K is the thermal conductivity of the medium,
  γ is the ratio of the molar heat capacities of the medium,
  C_v is molar heat capacity at constant volume,
  c is the velocity of sound in the medium

The combination of the two losses gives the total absorption for the system (classical absorption) as

$$\alpha_{c1} = \alpha_s + \alpha_{th} = \frac{2\pi^2 f^2}{\rho c^3}\left\{\frac{4}{3}\eta_s + \frac{(\gamma-1)K}{C_p}\right\} \qquad (4.9)$$

Since for any given liquid η, K, γ, ρ and C_p can be regarded as constant at constant temperature, the value α/f^2 should be independent of the experimental frequency employed to determine α. Experimentally this is not the case for many liquid systems, with α/f^2 decreasing with increase in frequency. This is due to the fact that the total energy content of a liquid is not restricted solely to translational energy, but is the sum of many components including rotational, vibrational, molecular conformational and structural forms. It is the coupling of the translational energy with these other energy forms which leads to the absorption of sound in excess of that deduced from eqn (4.9), and to the non-constancy of α/f^2 with increasing frequency. As will be seen later (Section 4.4), the measurement of α as a function of frequency will allow a determination of the various thermodynamic parameters associated with conformational change of the polymer. It must be noted however, that such determinations require the use of low-intensity sound waves, since it is only in the absence of cavitation that eqns (4.5–4.9) can be exact. In high-intensity fields the presence of cavitation bubbles alters the acoustic environment and the magnitudes of α, ρ and c will not be known accurately.

Although the various factors affecting the onset of cavitation have been discussed in detail in the introductory chapter, it is also useful to consider another simplistic explanation as to why cavitation bubbles occur at intensities substantially less than those predicted. It is well known that liquids boil when the pressure in the liquid is reduced to a value just below the liquid vapour pressure. (As a typical example, consider the process of evaporation of ether at a water pump.) If cavitation can be assumed to be a 'cold boiling' process, then cavitation bubbles will appear when the intensity (and hence P_A, eqn (4.6)) is increased sufficiently to ensure that at some instant (t) in the rarefaction cycle (eqn (4.2)) the total liquid pressure ($P_h - P_a$) is less than the vapour pressure, P_v, of the liquid. The consequence of this dependence on vapour pressure will be referred to later (Section 4.2.1).

In general there are two types of cavitation bubble—namely, transient and stable. Transient bubbles are those thought to be produced by intensities in excess of 10 W cm^{-2}, which on collapse give rise to very high temperatures and pressures[3] in accordance with eqns 4.10 and 4.11.

$$T_{max} = T_0 \left\{ \frac{P_m(\gamma - 1)}{P} \right\} \tag{4.10}$$

$$P_{max} = P \left\{ \frac{P_m(\gamma - 1)}{P} \right\}^{\gamma/(\gamma-1)} \tag{4.11}$$

Here T_0 is the ambient (experimental) temperature, γ is the ratio of specific heats of the gas (or gas–vapour mixture), P is the pressure in the bubble at its maximum size (usually assumed to be equal to the vapour pressure (P_v) of the liquid), and P_m is the pressure in the liquid at the moment of collapse, usually assumed to be ($P_h + P_A$). It is the existence of these very high temperatures within the bubble that have formed the basis for the explanation of radical production and hence polymerization, whilst the release of the pressure, P_{max}, as a shock wave, is a factor which has been used to account for polymer degradation.

The second type of bubble, stable bubbles, are believed to contain mainly gas and some vapour and are produced at fairly low intensities (1–3 W cm^{-2}), oscillating, often non-linearly, about some equilibrium size, for many acoustic cycles. The time scale over which they exist is sufficiently long that mass diffusion of gas, as well as thermal diffusion, with consequent evaporation and condensation of the vapour, can occur, resulting in significant long-term effects. As with transient cavitation, estimates have been made of the temperatures and pressures produced in the bubbles as they oscillate in resonance with the applied acoustic field. According to Griffing,[4] the ratio T_0/T_{max} is given by

$$\frac{T_0}{T_{max}} = \{1 + Q[(P_h/P_m)^{1/3\gamma} - 1]\}^{3(\gamma-1)} \tag{4.12}$$

where Q is a damping factor equal to the ratio of the resonance amplitude to the static amplitude of vibration of the bubble. Typically for a bubble containing a monatomic gas ($\gamma = 1\cdot67$) with $P_m/P_h = 3\cdot7$ (corresponding to an intensity of 2.3 W cm^{-2}) and assuming a value of $Q = 2\cdot5$, yields a value for T_{max} of 1665 K. Calculations of the local pressures due to these resonance vibrations has resulted in values which exceed the hydrostatic pressure by a factor of 150 000. It has been suggested that it is the presence of the intense local strains in the vicinity of the resonating bubble which is a contributing factor in the degradation of polymers in solution.

4.2 Degradation of polymers

Although it is now well established that the irreversible reduction in viscosity of polymer solutions subject to intense ultrasonic irradiation is as a result of cavitational bubble collapse, early workers[5,6] in the field ascribed the observations to the increased frictional forces set up between the macromolecule and the rapidly moving solvent molecules. Their conclusions were based upon observations that polymer degradation still occurred at high and low pressure (e.g. 15 atm[5] or vacuum[6]), conditions under which cavitation ought to have been totally negligible. Although subsequent work has proved these conclusions too simplistic, this early work was important in that it predicted

(a) that the higher the molar mass of the original macromolecule, the faster the degradation rate, and

(b) that there was a lower molar mass limit beyond which the macromolecule resisted further degradation.

Mathematically, observation (a) above may be expressed as

$$\frac{dx}{dt} = k(P_t - P_L) \tag{4.13}$$

where dx is the number of bonds broken in time dt, P_t is the degree of polymerization at time t and P_L is the limit degree of polymerization. Malhorta[7] has subsequently expressed bond breakage per unit time as

$$\frac{dx}{dt} = \frac{P_0/P_t - 1}{t} \tag{4.14}$$

where P_0 is the initial degree of polymerization ($\alpha \bar{M}_n$, the number-average molar mass of the polymer).

It is possible to semi-quantify observation (b) above in terms of a modified Stokes equation (eqn (4.15)), by assuming that a macromolecule consists of a number (n) of spherical entities, each of which is subject to a frictional interaction with the solvent:

$$F = 6\pi\eta rVn \qquad (4.15)$$

Here η is the fluid viscosity, r is the radius of the 'sphere', n is the number of 'spheres' or pendant groups on the polymer and V is the velocity. Using eqn (4.15), together with the spectroscopic estimate of the energy required to break a C–C bond (4.5×10^{-9} N) suggests a lower limit for the relative molecular mass (RMM) of polystyrene ($r \sim 3 \times 10^{-10}$ m) in benzene ($\eta = 6.2 \times 10^{-4}$ N s m^{-2}) of approximately 450 000 when employing ultrasound of intensity 5 W cm^{-2} ($V = 0.3$ m s^{-1}). However, in order to predict the effect of irradiation frequency, solvent, temperature, gas type and pressure, any discussion necessitates a consideration of cavitation and the factors involved.

4.2.1 Factors affecting polymer degradation

Both the creation and the collapse of cavitation bubbles require finite time periods. It has been argued that the higher the frequency of an ultrasonic source, the less likely that cavitation will occur and hence the less likely it is that the macromolecule will be degraded. In fact, Noltingk and Neppiras have argued that the frequency limit for cavitation is 1 MHz, and that above this frequency cavitation is impossible. Such a statement, however, does not find total support in the independent work of Gaertner[8] and Mostafa.[9] For example, although Mostafa has observed a maximum in the degradation rate for polystyrene in benzene at 1 MHz, there was still a discernible, albeit small, degradation at 2 MHz, the highest frequency employed. Here it must be assumed that cavitation is negligible. The explanation offered was that whereas only shear forces were operational at high frequency, a decrease in frequency led to a decrease in attenuation (eqn (4.9)) and subsequent increase in the real intensity in solution (eqn (4.5)), the accompanying cavitational effects of which increasingly complemented the shear forces. This appears to be substantiated by Gaertner who investigated depolymerization at both 400 kHz and 2.5 MHz and found that whereas degradation at the lower frequency only necessit-ated $0.5 \, \text{W cm}^{-2}$, the higher frequency required an intensity of approximately $2 \, \text{kW cm}^{-2}$ to obtain a comparable effect.

The choice of solvent and temperature are also important in maximizing the extent and rate of degradation. For example, eqn (4.11) implies that the maximum collapse pressure, P_{max}, should be inversely dependent upon the vapour pressure, P_v; i.e. an increase in P_v will lead to lower P_{max} values and hence lower degradation rates, a prediction which has been confirmed by many workers in the field. For example, Schmid and Rommel observed lower degradation rates in benzene ($P_v = 9.09 \, \text{kPa}$) than in toluene ($P_v = 3.35 \, \text{kPa}$) despite the fact that the viscosities of the two solvents were similar. Although similar trends have been observed for the degradation of nitrocellulose in a variety of esters as solvent, it should not be inferred that vapour pressure is the only contributing solvent parameter to degradation. For instance, whilst the increase in degradation rate of nitrocellulose in the solvents ethyl acetate, n-propyl acetate, i-butyl acetate and i-amyl acetate is certainly in accord with the decrease in solvent vapour pressure, it is also in accord with an increase in solvent viscosity. As has already been argued in Chapter 1, an increase in viscosity, whilst making it more difficult to cavitate the solvent, will, if cavitation does occur, result in greater cavitational collapse. Contrary to the idea of solvent vapour pressure, Doulah[10] has suggested that it is the solvent's solvating capacity which is of importance. The suggestion is that the energy released, as a shock wave, is more effective in 'good' solvents where the polymer chains are extended, than in 'poor' solvents where they are not. Such a suggestion is similar to that proposed by the early workers and implies frictional interaction.

Most workers[11-13] agree that degradation is increased in the pre-sence of gas. However, the extent and rate are dependent upon the gas type, its solubility and its thermal conductivity. For example, bubbles containing gas which have a large specific heat ratio, γ, should produce the largest shock waves on collapse (eqn (4.11)) and give rise to the largest rate and extent of degradation. This can be confirmed in part by considering Table 4.1 where the degradation rates, measured as $([\eta]_{t=0} - [\eta]_{t=20})/[\eta]_{t=0}$, for the first 20 min of irradiation, and the molar masses, measured as intrinsic viscosity, $[\eta]$ ($= KM^a$), after 60 min irradiation are given for solutions of polystyrene in benzene (1%) which have been saturated with a variety of different gases.

Although the diatomic gases (N_2, O_2, H_2, air) promote, as predicted,

Table 4.1 Degradation rates of polystyrene ($[\eta]_{t=0} = 1.80$) in benzene in the presence of various gases

	Gas							
	Air	N_2	O_2	H_2	*Ar*	NH_3	CO_2	SO_2
Rate/10^{-2}	47·5	43·0	42·5	40·0	33·7	12·7	8·3	5·0
$[\eta]_{t=60}$	0·68	0·80	0·80	0·68	0·77	1·06	1·36	1·72
γ	1·40	1·40	1·40	1·41	1·67	1·31	1·30	1·29
Solubility	0·14	0·11	0·22	0·07	0·24	9·95	2·4	88·0
$K/10^{-2}$ W m^{-1} K^{-1}	2·23	2·28	2·33	15·9	1·58	2·00	1·37	0·77

higher degradation rates, and lower extents of degradation than the polyatomic gases (NH_3, CO_2, SO_2), the rate of degradation in the presence of argon, a monatomic gas, is, surprisingly, lower than for the diatomics. Jellinek[14] has explained this anomaly in terms of the velocity at which the gas-filled cavities collapse. The larger the velocity of the cavity collapse, the faster the solvent molecules are swept past the polymer molecule and the faster is the degradation rate. An estimate of the average collapse velocity of a cavity when filled with the monatomic gas argon ($\gamma = 1.67$) reveals that it is approximately 70% of that when the cavity is filled with a diatomic gas such as oxygen ($\gamma = 1.4$), a ratio which is similar to the observed degradation rate—i.e. $33.7/42.5 = 80\%$. If the analysis is applied to the polyatomic gases, it is expected that, because of their lower γ values, they would have even faster degradation rates still, although the differences ought to be smaller than between the monatomics and diatomics. Consideration of Table 4.1 shows that this is not observed in practice and the polyatomics are found to have considerably lower rates. In fact, unlike the diatomics, the degradation rates for the polyatomics are also quite dissimilar, with the rates generally decreasing with increase in solubility. The conclusion must be that during the compression cycle when the pressure in the system is a maximum, gases with large solubilities do not undergo violent cavitational collapse but simply redissolve.

However, if solubility were the only other contributing factor to degradation, a solution saturated with hydrogen ought to degrade faster and to a greater extent than one saturated with any of the other diatomic gases. Since this does not occur, it must be assumed there are other factors of importance—e.g. thermal conductivity (K) of the gas. For example, if it can be assumed that a gas with a high thermal conductivity value is more effective in dissipating the heat formed in the bubble during collapse and thereby lowering the maximum temperature, T_{max}, attainable in the bubble, this might explain the slightly lower degradation rates in the presence of hydrogen ($K = 15.9$) than the other diatomics ($K \sim 2.3$).

In the introduction to this chapter it was suggested that cavitation would only occur at that instant in the rarefaction cycle when the liquid pressure ($\sim P_h - P_a$) fell below the vapour pressure of the liquid, and as such depended indirectly on the intensity of the sound wave (eqns (4.2) and (4.6)). Thus, if degradation is a result of cavitation it ought only to occur when the cavitation threshold is exceeded ($P_v \sim P_h - P_a$) and to increase with increase in intensity (eqns (4.5) and (4.11)). The limiting RMM ought also to decrease with increase in intensity since the bubble collapse velocity will be larger (i.e. $V = P_a/\rho c$). (For example, eqn (4.15) suggests that the degradation of polystyrene in benzene at intensities of 7 and $26.5\ W\ cm^{-2}$ should produce limits of 380 000 and 190 000, respectively). Both increase in rate and lowering of RMM with increased intensity have been observed experimentally.[11,13,15,16] However, it must be recognized that there is an optimum power which can be applied to a system to obtain the most beneficial effect. Qualitatively, it may be suggested that the bubbles produced at very high intensities serve to reduce power dissipation by reflecting the sound wave and creating a non-linear response to the increase in intensity. Quantitatively, it can be argued that at high intensities (i.e. large P_A values) the cavitation bubbles have grown so large on rarefaction (R_{max}, eqn (4.16)) that there is insufficient time available for collapse during the compression cycle.

$$T = 0.915 R_{max} (\rho/P_m)^{1/2} \tag{4.16}$$

(Here T is the collapse time for a bubble in a liquid of density ρ and is assumed to be less than one-fifth of the compression cycle period.)

The effect of pressure on ultrasonic degradation has led to many contradictory results. Whereas Schmid & Rommel[5] observed that 5 and 8 atm, respectively, of excess pressure slowed and eventually stopped the ultrasonic degradation of nitrocellulose, the application of higher excess pressures, 8 and 15 atm, only resulted in a slowing down of the degradation rate for styrene, and then to the same extent for both pressures. In contrast, Mark[6] has observed *increases* in the degradation rate with increases in pressure for a similar system.

Those workers who have investigated the degradation of polystyrene as a function of the initial molar mass of the sample have all found

   (a) that the degradation rate was highest for the sample with the
       largest RMM, and
   (b) that the curves for all samples converged in the later stages,
       apparently reaching the same final relative molar mass.

Similar relationships have also been observed for the degradation of
poly(vinyl acetate), poly(methyl acrylate) and nitrocellulose. The
explanation most usually offered for such observations is that the
larger the RMM of the macromolecule the less likely it is to move
bodily in the alternating ultrasonic field and the statistically more
likely it is that the solvent molecules, as they are swept past the
macromolecule, will interact with a segment of the polymer and
rupture a C–C bond.

   In investigations of the effect of solution concentration on degradation, the rate and extent of degradation have been observed to
decrease with increase in concentration. These observations have been
interpreted in terms of the increase in viscosity of the solution—i.e.
the higher the viscosity the more difficult it becomes to cavitate the
solution, at a given intensity, and the smaller is the degradation effect.
This obviously has important consequences in a polymerization
process conducted in the presence of ultrasound, since as the polymer
concentration, and hence viscosity, increases with time, the acoustic
environment of the system will change.

   In general, maximum degradation rates and largest extents of
degradation can be obtained by:

   (a) saturating the polmer solution with a gas;
   (b) employing a gas with a low solubility;
   (c) using a solvent with a low vapour pressure;
   (d) reducing the experimental temperature;
   (e) reducing the insonation frequency;
   (f) increasing the intensity of irradiation;
   (g) decreasing the solution concentration;
   (h) increasing the RMM of the polymer.

4.2.2   Degradation mechanisms

Currently there are three proposed models to account for polymer
breakdown in the presence of an ultrasonic field. We have already
discussed the first, the Jellinek model,[14] in which Schmid's ideas[5] of

increased frictional force generated on cavitational collapse were used to explain the effects of various gases (Table 4.1).

The second model is mathematical in origin and is credited to Doulah.[10] In it he suggests that the released shock-wave energy on bubble collapse gives rise to a series of eddies which interact with the macromolecule in solution. By calculating the dynamic force set up across the whole length of the polymer chain, Doulah has been able to deduce that not only is the degradation rate dependent upon the acoustic intensity and initial size of the macromolecule, but that there is a minimum chain length below which degradation cannot occur; these predictions have been confirmed experimentally.

The final model is that proposed by Glynn,[17] which unlike the previous two predicts the point, or points, within the macromolecule where bond cleavage is likely to occur. Although it might be assumed that cleavage would occur at points of inherent weakness within the polymer backbone, as is the case in chemical or photodegradation, work by Melville[13] seems to refute this, as does Glynn's prediction that cleavage is Gaussian about the mid-point of the chain.

Wherever the site of cleavage, the result must be the production of macroradicals,[18] the existence of which have been confirmed spectroscopically by the use of radical scavengers such as diphenyl picrylhydracyl (DPPH). Obviously, in the absence of scavengers the macroradicals are free to combine by either disproportionation or combination termination, the former leading to smaller-sized macromolecules whilst the latter will give a distribution dependent upon the size of the combining fragments. Since it has already been suggested that there is a lower RMM limit below which macromolecules appear to be unaffected by ultrasound, it must be anticipated that ultimately there will be a lowering of both RMM and heterogeneity ($HI = \bar{M}_w / \bar{M}_n$) with the passage of time. Although such predictions have been confirmed experimentally by several workers, recent work by Lorimer *et al.*[19] with a bimodal polymer sample (Figs 4.1 and 4.2) has shown an initial increase in $\bar{M}_n$ with irradiation (Table 4.2). This observation has been rationalized in terms of a higher probability and more rapid degradation of the larger—RMM sample (peak B, Fig. 4.2) giving rise to molecules with a higher RMM than the average for peak A (i.e. increase in $\bar{M}_n = \Sigma N_i M_i / \Sigma N_i$) but lower than average RMM for peak B (i.e. lower $\bar{M}_w = \Sigma W_i M_i / \Sigma W_i$). This observation has been noted for both a commercially obtained sample and samples produced experimentally by a free-radical mechanism.

Table 4.2 Degradation of poly(N-vinylcarbazole) (PNVK) in
benzene

	Irradiation time (min)					
	0	10	20	30	60	120
$\bar{M}_n/10^3$	296	326	245	245	183	143
$\bar{M}_w/10^3$	715	555	415	398	276	211
HI	2·42	1·71	1·69	1·63	1·51	1·48

Although most of the reported literature seems to infer that
degradation of a macromolecule takes place by main-chain scission,
one should not rule out the possibility of C—H scission, since the bond
energy for the latter ($412\,kJ\,mol^{-1}$) is only some 20% higher than that
for the former ($350\,kJ\,mol^{-1}$). The consequence of such a scission
would on termination by combination lead to branched structures. The
suggestion that such a mechanism may be possible appears to be
confirmed by recent degradation studies of Lorimer et al.[19] (Table
4.3).

In general, k' is usually found to lie in the range 0·3–0·5 for a
randomly coiling molecule and to increase with increase in molar
mass. Although higher values than 0·5 have been taken to indicate
branching within the polymer, i.e. $k' = 0.6$ for flexible molecules, 0·77

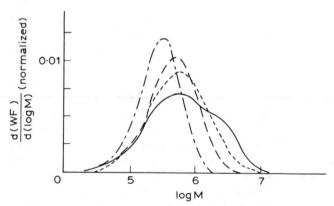

Fig. 4.1 RMM distribution function: PNKV in dichloromethane. Irradiation
times: solid curve, 0 min; dotted curve, 5 min; dashed curve, 15 min; chain
curve, 30 min. (WF = weight fraction.)

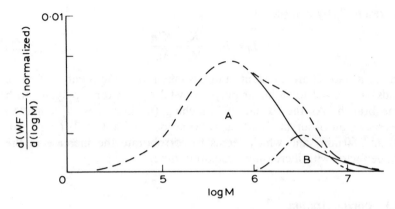

Fig. 4.2 RMM distribution function: PNVK in dichloromethane (see text). (WF = weight fraction.)

for rigid rods and 2–2·2 for spherical molecules,[20] this is not a view shared by all workers. A more universally accepted criterion for branching is that the intrinsic viscosity, $[\eta]$, for a branched polymer should be lower than that for a linear polymer of the same $\bar{M}_n$. Where comparisons of $[\eta]$ at comparable $\bar{M}_n$ are possible (e.g. $t = 20$ and 30 min, $\bar{M}_n = 245\,000$), the indication is that branching is occurring, since the $[\eta]$ values decrease with irradiation time.

If branching occurs within a given polymer then it is usual to

Table 4.3 Degradation of PNVK in benzene

	Irradiation time (min)					
	0	*10*	*20*	*30*	*60*	*120*
$[\eta]/10^{-2}\,\mathrm{dm^3\,g^{-1}}$ [a]	6·76	6·08	4·92	4·24	3·52	3·56
k' [b]	0·62	0·63	0·90	1·34	1·84	0·52
T_g/K [c]	500	503	502	501	497	503
$M_n/10^3$	296	326	245	245	183	143

[a] $[\eta]$ is the limiting viscosity, obtained from the intercept of the η_{sp}/c vs c plot.
[b] k' is the slope of the η_{sp}/c vs c plot.
[c] T_g is the glass transition temperature.

represent T_g by eqn (4.17):

$$T_g = T_g^0 - \frac{K}{\bar{M}_n} - \frac{K'n}{\bar{M}_n} \qquad (4.17)$$

where K and K' are constants and n is taken to be the number of chain ends—i.e. $n = 2$ for a linear polymer, whilst $n = 3$ for a polymer with one branch. As the extent of branching (n) increases, T_g ought to decrease, an observation which is borne out in Table 4.3 ($t = 10$, 20, 30 and 60 min) and which seems to corroborate the increase in the shape factor with increasing irradiation time.

4.3 Polymerization

In general, sonochemical polymerizations can be thought to fall into one of four categories:

(1) application of ultrasound to solutions containing two homopolymers;

(2) application of ultrasound to a solution containing a polymer and a monomer (the monomer may or may not be the same as that contained in the polymer);

(3) application of ultrasound to a solution containing only monomer;

(4) application of ultrasound to a solution containing monomer and initiator.

Of the various attempts which have been made to produce block or graft copolymers using a combination of homopolymers, the success has been somewhat mixed. For example, Henglein[21] has successfully produced both graft and block copolymers using polystyrene and poly(methyl methacrylate), and Keqiang[16] has produced water-soluble block copolymers of hydroxycellulose and polyethylene oxide. Malhorta[7] on the other hand, employing a variety of homopolymers (rigid and flexible) met with only limited success. In effect those homopolymers with bulky substituents, although able to undergo chain scission did not result in scrambled copolymers in the presence of polystyrene.

Berlin[22] has produced a block copolymer of poly(methyl methacrylate) with acrylonitrile monomer, and confirmed the prediction that the time required to produce a given amount of poly(acrylonitrile) in the block decreases with increasing intensity. Keqiang, also working with acrylonitrile, has been able to produce both water-soluble and water-insoluble block copolymers, depending on the initial acrylonitrile charge and irradiation time employed. Driscoll[23] has studied the homopolymerization of the monomers styrene and methyl methacrylate in the presence of their respective homompolymers and observed that the lower the reaction temperature the faster was the reaction rate and the higher the final polymer yield. (Again, this confirms the prediction that the lower the vapour pressure, P_v, the larger is the bubble temperature (eqn (4.10)) and pressure (eqn (4.11)) on collapse, and thus the greater is the shock wave produced.) Melville,[13] studying the ultrasonically induced polymerization of styrene, methyl methacrylate and vinyl acetate in the presence and absence of poly(methyl methacrylate) found that the polymerization rates (~1% conversion/h) were not substantially increased by the presence of polymer. The conclusion was that the degradation of poly(methyl methacrylate) was not the major source of radical production and dispels the view, held by previous workers, that a polymerization reaction could only occur in the presence of polymer. Using hydroquinone as an inhibitor Melville was able to deduce that the rate of radical production from ultrasound was $\sim 2 \times 10^{-9} \, mol \, dm^{-3} \, s^{-1}$.

Although there have been several kinetic studies of ultrasonically induced polymerization in the absence of polymer, there are only a few[19,24,25] investigations into the effect of the various reaction parameters, such as irradiation intensity and frequency, reaction temperature, solvent vapour pressure, etc.

One of the first observations of the sonochemical initiation of polymerization of a monomer was that by Lindstrom and Lamm in 1950[24], in which they investigated the polymerization of acrylonitrile in dilute aqueous solution at several different acoustic frequencies and intensities. By assuming that the initiating species were either H· or OH· radicals (or both), produced during the cavitation process, or that radicals were produced as a result of polymer (P) degradation, the authors proposed the following mechanism

$$H_2O \xrightarrow{k_1} H\cdot + OH\cdot \ (= 2R\cdot) \qquad (4.18)$$

$$P \xrightarrow{k_2} 2R\cdot \qquad (4.19)$$

$$R\cdot + nM \xrightarrow{k_3} R_n\cdot \qquad (4.20)$$

$$R_n\cdot + R_m\cdot \xrightarrow{k_4} \text{dead polymer} \qquad (4.21)$$

On applying the steady-state approximation, they deduced that the radical concentration, $[R\cdot]$, and propagation rate, R_p, were given respectively by

$$[R\cdot] = [(k_1 + k_2[P])/k_4]^{1/2} \qquad (4.22)$$

and

$$R_p = -\frac{d[M]}{dt} = k_3[M][R\cdot] \qquad (4.23)$$

On expressing both the monomer and polymer concentrations in terms of p ($= ([M]_0 - [M])/[M]_0$), the fraction polymerized (or extent of reaction), eqn (4.23) reduces to

$$\frac{dp}{dt} = (1-p)k_3\left\{\frac{(k_1 + [M]_0 k_2 p/n)}{k_4}\right\}^{1/2} \qquad (4.24)$$

where n is the average number of monomers in the polymer, i.e. $[P] = ([M]_0 - [M])/n$. For this particular polymerization reaction, the extent of reaction over the time interval investigated never exceeded 0.02. By making the assumption that $(1-p)$ could be approximated to 1, integration of eqn (4.24) yielded eqn (4.25), a relationship which the authors found fitted the data extremely well:

$$t = c_1[(1 + c_2 p)^{1/2} - 1] \qquad (4.25)$$

where

$$c_1 = \frac{2nk_1^{1/2}k_4^{1/2}}{k_2 k_3 [M]_0} \quad \text{and} \quad c_2 = \frac{k_2[M]_0}{k_1 n}$$

The experiments by Kruus[25] with bulk methyl methacrylate (MMA) initiated by ultrasound have provided polymer yields which are too small ($<4\%$/h) to suggest that the technique can provide a commercial alternative to the conventional methods. Nevertheless, this is an important investigation in that it is one of the rare systematic studies into the effects of acoustic intensity, temperature and reaction volume. To explain the observed variations of polymerization rate R_p with the various parameters (i.e. $R_p \propto (\text{intensity})^{1/2}$, $R_p \propto 1/\text{volume}$), Kruus adopted a mechanism similar to that of Lindstrom and Lamm[24] with the exceptions that the cavitation bubbles were regarded as a reactant and that for low conversions very few initiating radicals were produced via degradation of the polymer.

$$M + C \xrightarrow{k_1} 2R\cdot \qquad (4.26)$$

$$P + C \xrightarrow{k_2} 2R_n\cdot \qquad (4.27)$$

$$R\cdot + nM \xrightarrow{k_3} R_n\cdot \qquad (4.28)$$

$$R_n\cdot + R_m\cdot \xrightarrow{k_4} \text{dead polymer } (P_{2n} \text{ or } P_{n+m}) \qquad (4.29)$$

$$P_{2n} + C \xrightarrow{k_5} 2R_n\cdot \qquad (4.30)$$

By applying the steady-state analysis and assuming the concentration of cavitation bubbles, $[C]$, could be expressed by eqn (4.31), Kruus deduced that the propagation rate could be expressed by eqn (4.32):

$$[C] = \frac{N_c}{V} = \frac{FI}{V} \qquad (4.31)$$

(N_c is the number of cavitation bubbles, V is the reaction volume, I is the acoustic intensity and F is a proportionality constant.)

$$R_p = -\frac{d[M]}{dt} = k_3[M][R\cdot] = K[M]^{3/2}\left(\frac{FI}{V}\right)^{1/2} \qquad (4.32)$$

where K is a constant $= (2fk_1/k_4)^{1/2}k_3$, and f is an efficiency factor introduced to represent the fraction of radicals which actually initiate the polymerization.

For polymerizations performed at constant intensity ($20\ \text{W cm}^{-2}$) and variable temperature ($-18°\text{C}$ to $40°\text{C}$), Kruus obtained an activation energy of $19.3\ \text{kJ mol}^{-1}$, a value which was in good agreement with that reported ($18\text{--}20\ \text{kJ mol}^{-1}$) for the bulk polymerization of methyl methacrylate if the contribution from initiation was excluded. This suggests that initiation by ultrasound is independent of temperature or has neglible activation energy, a well-known characteristic of photopolymerization, as is the R_p vs $[\text{intensity}]^{1/2}$ dependence.

Kruus also conducted experiments in the presence of the radical scavenger DPPH and observed induction periods which were roughly proportional to the concentration of DPPH employed. This clearly demonstrates the free-radical nature of the polymerization. By assuming that each of the monomer radicals produced by the cavitation process (eqn (4.26)) reacted with one DPPH molecule, he was able to deduce the following kinetic relationship:

$$-\frac{d[DPPH]}{dt} = 2fk_1[M]_0\left(\frac{FI}{V}\right) \qquad (4.33)$$

which on integration yields

$$[DPPH]_0 - [DPPH]_t = 2fk_1[M]_0\left(\frac{FI}{V}\right)t \qquad (4.34)$$

Excellent correlations were obtained (better than 0·97) for plots of the consumed DPPH with time, intensity and reciprocal volume.

Experiments were also performed in the temperature range −18°C to 40°C in the presence of DPPH. Although the data fitted the conventional Arrhenius relationship (eqn (4.35)), it gave an effective activation energy which was *negative* ($-85\,kJ\,mol^{-1}$),

$$\ln k = \ln A - \frac{E_a}{RT_0} \qquad (4.35)$$

This observation prompted Kruus to assume that

(a) radical formation took place in the bubble;
(b) the bubble reaction mechanism was governed by the usual Arrhenius behaviour (eqn (4.35)), and
(c) the reaction temperature was in fact the bubble temperature, T_m (eqn (4.10));

and to fit the data (i.e. $\ln k$ vs P_v/T) to a modified Arrhenius equation (eqn (4.36)):

$$\ln k = \ln A - \frac{E_a P_v}{RT_0(\gamma - 1)P_m} \qquad (4.36)$$

Although Kruus makes little further comment except that an excellent correlation was observed, it is interesting to note that if it can be assumed that the cavitation bubbles are filled mainly with argon gas ($\gamma = 1·67$) (the experiments were conducted with a gas flow rate of $20\,cm^3\,s^{-1}$), and that the pressure on bubble collapse, P_m, ($= P_A + P_h$) is approximately 8 atm ($P_A^2 = 2I\rho c$; eqn (4.6)), the slope of $\ln k$ vs P_v/T_0 (eqn (4.36)) provides an estimate for E_a of $460\,kJ\,mol^{-1}$, a value close to that for the bond energy of a C—C bond ($345\,kJ\,mol^{-1}$). Obviously the presence of methyl methacrylate ($\gamma = 1·05^{26}$) will lower the value of γ for the 'gas' resident in the bubble, thereby lowering the effective E_a. However, the generality of the analysis, even allowing for the assumptions, would indicate that the site of radical formation is indeed the cavitation bubble itself.

Miyata and Nakashio[27] have studied the effect of frequency and intensity on the thermally initiated (2,2-azobisisobutyronitrile, AIBN) bulk polymerization of styrene and found that whereas the mechanism of polymerization was not affected by the ultrasound, the overall rate constant, K $(R_p = K[M])$ *decreased* linearly with increase in the intensity. They interpreted the decrease in the overall value of K as being caused by either an increase in the termination reaction (i.e. specifically the termination rate constant, k_t) or a decrease in the initiator efficiency. The increase in $k_t = (k_t^0/\eta)$ is the more reasonable, in that ultrasound is known to reduce the viscosity of polymer solutions. Unfortunately, little can be learned from Myata's and Nakashio's work at different frequencies. Employing a constant output power from the sound generator, and operating at 200, 400, 600 and 800 kHz, they observed a maximum for the value of K at 400 kHz. Whether the maximum is meaningful is uncertain because of the experimental conditions employed. It has already been suggested that larger input intensities are necessary, at the higher frequencies, if equivalent sonochemical effects are to be produced.

Lorimer *et al.*[19] have compared the effect of ultrasound on the solution polymerization of *N*-vinylcarbazole (NVC) in both the presence and absence of the same initiator (AIBN). They have found that in the presence of initiator at a given temperature and irradiation frequency ($T = 60°C$, $f = 20$ kHz) there was a maximum in the R_p vs I curve. Moreover, they also observed that

(1) at a given irradiation power there was an optimum irradiation time beyond which there was a decrease in the amount of polymer produced;

(2) at lower temperatures and intensities the optimum polymer yields were increased (Fig. 4.3).

In the absence of the initiator, but in the presence of very high-intensity ultrasound ($I = 500$ W cm^{-2}), polymerization was still observed to occur and produced 10% polymer in approximately 30 min (Table 4.4). The molar masses, however, were significantly larger than those produced by the conventional solution polymerization (Table 4.5).

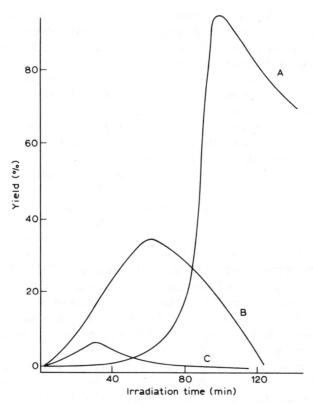

Fig. 4.3 Effect of ultrasound on the polymerization of *N*-vinylcarbazole. A, $T = 60°C$, $I = 27·6 \text{ W cm}^{-2}$. B, $T = 50°C$, $I = 59·6 \text{ W cm}^{-2}$. C, $T = 60°C$, $I = 59·6 \text{ W cm}^{-2}$. (Reproduced with permission from Mason, T. J. & Lorimer, J. P. (eds) (1988). *Sonochemistry: Theory, Application and Uses of Ultrasound in Chemistry*. Ellis Horwood, Chichester.)

Table 4.4 Sonochemical polymerization of NVC in the absence of initiator ($[M] = 1 \text{ mol dm}^{-3}$)

Temperature range (°C)	Volume (cm³)	Reaction time (min)	Yield (%)	$\bar{M}_n$ ($\times 10^{-3}$)	$\bar{M}_w$ ($\times 10^{-3}$)	HI	k'
61–72	25	30	11	1150	2600	2·3	2·1
58–78	50	60	10	1100	2200	2·0	1·5

Since, mathematically, the rate of production (R_p) and the number average molar mass of polymer may be represented by eqns (4.37) and (4.38), respectively, then for two polymerization reactions differing only in the concentration of initiator employed, eqns (4.39) and (4.40) are more representative. In other words, a fourfold decrease in the initiator concentration should lead to a twofold decrease in rate (or yield per unit time) and a twofold increase in $\bar{M}_n$ (Table 4.5):

$$R_p = k_p[\text{M}]\left(\frac{2fk_d[\text{I}]}{k_t}\right)^{1/2} \tag{4.37}$$

$$\bar{M}_n = \frac{k_p[\text{M}]M_0}{(2fk_dk_t[\text{I}])^{1/2}} \tag{4.38}$$

$$R_p = K[\text{I}]^{1/2} \tag{4.39}$$

$$\bar{M}_n = \frac{K'}{[I]^{1/2}} \tag{4.40}$$

(Here k_d, k_p and k_t are the initiation, propagation and termination rate constants, [M] and [I] are the monomer and initiator concentrations and M_0 is the molar mass of the monomer).

A comparison of the values in Table 4.4 with those in Table 4.5, together with eqns (4.39) and (4.40), would suggest that ultrasonic initiation is producing the same radical concentration as would be produced thermally from an initiator whose concentration was 6×10^{-4} mol dm^{-3}.

On application of ultrasound to the system containing initiator, several points may be observed (Table 4.6). Firstly, the RMM values are lower than in the absence of ultrasound (Table 4.5) and decrease with irradiation time as expected. This suggests that the polymer is being degraded as it is produced. Secondly, the yield of polymer per unit time is greater than that produced thermally (Table 4.5, yield = 30% at $t = 30$). This is not unexpected, since fragmentation of polymer

Table 4.5 Thermal polymerization of NVC ([M] = 1 mol dm^{-3}) in benzene at 60°C

$[I]_0$ ($\times 10^3$)	Reaction time (min)	Yield (%)	$\bar{M}_n$ ($\times 10^{-3}$)	$\bar{M}_w$ ($\times 10^{-3}$)	HI	k'
10	30	31	285	700	2·5	0·7
2·5	30	13	530	1300	2·5	—

Table 4.6 Sonochemical polymerization $(I = 500 \text{ W cm}^{-2})$ of NVC $([M] = 1 \text{ mol dm}^{-3})$ in the presence of initiator $([I] = 10^{-2} \text{ mol dm}^{-3})$.

Temperature range (°C)	Reaction time (min)	Yield (%)	M_n $(\times 10^{-3})$	M_w $(\times 10^{-3})$	HI	k'
54–69	8	18	281	671	2·4	2·0
63–79	12	32	246	566	2·3	0·9
58–69	18	45	164	385	2·3	0·7

by the ultrasound will provide an additional radical source to that provided by thermolysis of the initiator. Thirdly, the apparent shape (k' value) is changing with irradiation time.

By deducing the average polymer yields which would have been obtained in a thermally polymerizable system at the average of the three sonochemical–thermal polymerization temperatures, the authors suggested that ultrasound is providing for a doubling of reaction rate, or approximately a 1% increase in polymer yield per minute of irradiation. Although it is reasonable to assume that the increase in rate is as a result of an increase in radical concentration, they do not

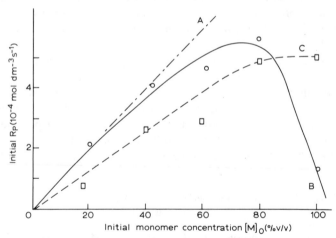

Fig. 4.4 Effect of ultrasound on the aqueous polymerization of N-vinylpyrrolidone. R_p vs $[M]_0$ curves: A, theoretical; B, without ultrasound; C, with ultrasound. (Reproduced with permission from Mason, T. J. & Lorimer, J. P. (eds) (1988). *Sonochemistry: Theory, Application and Uses of Ultrasound in Chemistry*. Ellis Horwood, Chichester.)

state categorically whether this is due to the degradation of the polymer, solvent or initiator. However there is no doubt that radicals are produced from the solvent and/or monomer as shown by Table 4.4.

A comparison of the effect of irradiation volume (Table 4.4) indicates that doubling the reaction volume requires twice the irradiation period. However, the longer irradiation time has appeared to have led to decreases in the RMM, *HI* and shape factor as expected.

Lorimer and Mason[28] have also investigated the effect of ultrasound on the aqueous polymerization of *N*-vinyl pyrrolidinone (NVP). This particular monomer does not follow the normal rate (R_p) monomer [M] dependence $(R_p = K[M]$, Fig. 4.4 curve A), but exhibits a maximum in the rate at 80% monomer (v/v) of monomer to water Fig. 4.4 curve B). This R_p maximum is thought, from viscosity and enthalpy of mixing studies (Fig. 4.5), to be due to the formation of a hydrogen-bonded monomer–water complex. Previous investigations

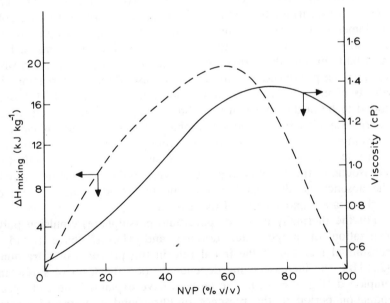

Fig. 4.5 Heat of mixing and viscosity for aqueous *N*-vinylpyrrolidone. (Reproduced with permission from Mason, T. J. & Lorimer, J. P. (eds) (1988). *Sonochemistry: Theory, Application and Uses of Ultrasound in Chemistry*. Ellis Horwood, Chichester.)

by the authors of the solvolysis of 2-chloro-2-methylpropane had led them to conclude that low-intensity ultrasound ($<2 \, W \, cm^{-2}$) was capable of destroying hydrogen-bonds, and if this were so then application of ultrasound to the NVP system ought to lead to a destruction of the complex and a decrease in R_p, Except for the pure monomer system (where there could not be hydrogen-bonding, since water was absent) all R_p values were reduced in the presence of ultrasound (Fig. 4.4 curve C). It can only be assumed that some other effect, perhaps chemical rather than physical, is operating in the pure monomer system since the R_p value increased in the presence of ultrasound. Possible explanations are that the application of ultrasound leads either to increased breakdown of the initiator, an observation which has been verified independently, or to the production of monomer radicals, as suggested by Kruus.

Ultrasonic waves have also been found to increase the rates of emulsion and suspension polymerizations. For example, Hatate[29] has investigated the suspension polymerization of styrene under ultrasonic irradiation in both batch and continuously stirred reactors and observed that irradiation was an effective method of preventing both agglomeration between the droplets and the sticking of droplets to the reactor wall. Both these factors can lead to serious problems such as heat build-up and the formation of large masses making it impossible to carry out prolonged operations. In the case of the batch reactor, the effects of ultrasound on the conversion and the average RMM were found to be negligibly small and Hatate suggested that the ultrasonic energy supplied to the monomer phase (500 W), whilst sufficient to prevent agglomeration, was not great enough to effect the polymerization rate. Unfortunately, is was not possible to quantify the effect of ultrasound in the continuous process owing to the extreme difficulty in the absence of ultrasound in realizing the steady state because of agglomeration and sticking of the droplets.

For the thermally initiated (potassium persulphate) emulsion polymerization of polystyrene, Lorimer and Mason[30] have found a substantial increase in the initial rate in the presence of ultrasound (20 kHz), the increase being dependent upon the level of surfactant employed (Figs 4.6–4.8). The authors have explained the lack of an induction period in the presence of ultrasound in terms of greater radical production induced by sonication, i.e. enhanced initiator breakdown or polymer degradation, and the creation of a far more stable emulsion. Although the latter point was confirmed visually, the

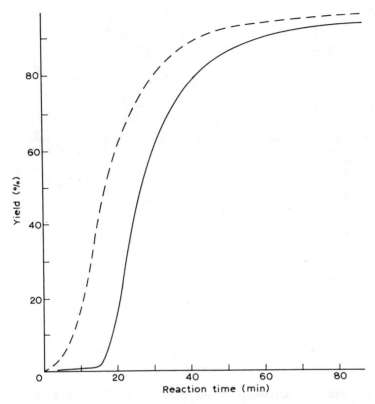

Fig. 4.6 Effect of ultrasound on polymer yield $[(E) = 8.25 \text{ g kg}^{-1}]$: solid curve, conventional; broken curve, ultrasonic.

acceleration was thought most likely to be due to increased radical production, since they observed increased initiator breakdown (40%) and decreased molar mass in the presence of ultrasound. (RMM values fell from 12×10^6 to 1×10^6 depending upon the irradiation power and surfactant level chosen.) Since irradiation only appeared effective during the initial part of the reaction (<25 min), the authors applied ultrasound for an initial 30-min period before allowing the reaction to proceed conventionally (Fig. 4.9). The result was an improved yield in the early period of the reaction when compared to the conventional polymerisation, followed by a slightly improved yield in the latter stages, when compared to the continuously sonicated reaction.

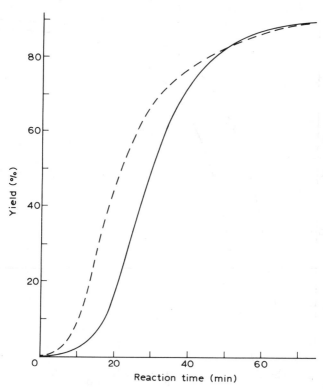

Fig. 4.7 Effect of ultrasound on polymer yield ($[E] = 4.15\,\mathrm{g\,kg^{-1}}$): solid curve, conventional; broken curve, ultrasonic.

Toppare[31] has investigated the effect of ultrasound (25 kHz bath) on both the polymerization rate and composition of the copolymers produced by the electro-initiated cationic polymerization of isoprene with α-methylstyrene. In the absence of ultrasound, the yield of copolymer was found to decrease with increase in the applied polymerization potential owing to the formation at the electrode of a polymer film. These films created a resistance to the passage of current into the bulk medium with consequent reduction in rate and yield. In the presence of ultrasound, however, the total conversion was found to exhibit a slight increase with increase in the polymerization potential (E_{pol}). This increase was attributed to a 'sweeping clean' of

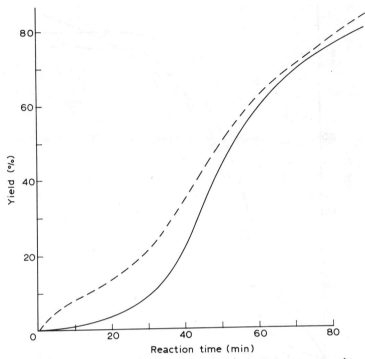

Fig. 4.8 Effect of ultrasound on polymer yield ($[E] = 2 \cdot 10\,\mathrm{g\,kg^{-1}}$): solid curve, conventional; broken curve, ultrasonic.

the electrode surface by the ultrasound. The authors also noted that both the proportion of isoprene incorporated into the polymer and the reactivity ratios (Table 4.7) were affected by the polymerization potential.

Ultrasonic irradiation has also had an influence on the properties of films produced by the electrochemical polymerization of thiophene. In the absence of ultrasonic irradiation, the films gradually became brittle as the electrolytic current density exceeded $5\,\mathrm{mA\,cm^{-2}}$. In contrast, flexible and tough films (tensile modulus, $3 \cdot 2\,\mathrm{GPa}$, and strength $90\,\mathrm{MPa}$) were obtained even at high current density ($10\,\mathrm{mA\,cm^{-2}}$) in the presence of ultrasound.[32] Results from cyclic voltammetry imply, not surprisingly, higher diffusion rates for the films prepared under the action of the ultrasonic field.

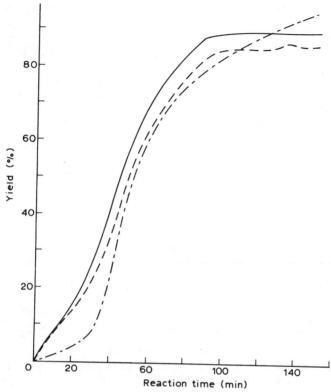

Fig. 4.9 Effect of irradiation time on polymer yield. Irradiation times: chain curve, 0 min; solid curve, 30 min; broken curve, 180 min.

Table 4.7 Effect of polymerization potential on reactivity ratio

E_{pol} (V)	Reactivity ratio		$[\eta]$
	Isoprene	α-Methylstyrene	
2·4	0·26(0·81)	0·19(0·78)	0·112(0·049)
2·6	0·36(0·77)	0·23(0·77)	0·110(0·057)
2·8	0·48(0·70)	0·50(0·98)	0·098(0·045)
3·0	0·80(0·97)	0·75(0·97)	0·044(0·050)

Figures in parentheses are in the presence of ultrasound.

4.4 Relaxation phenomena

In the introduction to this chapter it was suggested that whilst the absorption (attenuation) coefficient, α, for any system ought to be constant (eqn (4.9)) at constant temperature, experimentally this was not the case and α/f^2 was observed to vary with frequency. This is due to the fact that the total energy content of a medium is not restricted solely to translational energy, but is the sum of many components. It is the coupling of the translational energy with these other energy forms which leads to the absorption of sound in excess of that deduced from eqn (4.9) and to the non-constancy of α/f^2 with increasing frequency. The occurrence of this excess absorption is most easily illustrated by considering the fate of a vibrationally excited molecule, produced as a result of the energy interchange between the translation and vibrational modes. Provided the vibrationally excited molecules can be deactivated (by inelastic collisions with other molecules) and returned to the ground state in a time period which is shorter than the period of sound oscillation, the energy will be returned to the system in phase with the sound wave, and no net loss will be observed in the sound energy per cycle. As the frequency of the sound wave increases (i.e. as the time period decreases) the return of energy will become increasingly out of phase with the wave and will appear as an energy loss. Ultimately if the period of the wave is decreased sufficiently (i.e. if very high frequency ultrasound is applied) a situation will be reached where the perturbation of translational energy occurs so fast that there is no time available for exchange with the other energy forms. Between these two extremes of high and low frequency there exists a condition in which the frequency of the fluctuation induced by the sound wave is comparable with the time required for the energy exchange. The time lag between the excitation and de-excitation processes is observed as an acoustic relaxation. Since detailed discussions of the theoretical basis of acoustic relaxation have been published elsewhere[33,34] it is sufficient to report here that any relaxation is observable as an increase in the velocity–frequency curve, a peak in the μ–frequency curve, or a decrease[35,36] in the α/f^2–frequency curve (Fig. 4.10).

For the last case, the experimental data may be represented by eqn (4.41)

$$\frac{\alpha}{f^2} = \frac{A}{1 + (f/f_r)^2} + B \tag{4.41}$$

144   J. P. Lorimer

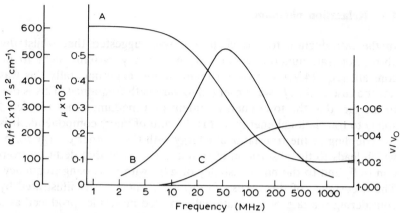

Fig. 4.10 Variation of ultrasonic parameters with frequency for a single relaxation process: A, α/f^2; B, μ; C, sound velocity. (Reproduced with permission from Mason, T. J. & Lorimer, J. P. (eds) (1988). *Sonochemistry: Theory, Application and Uses of Ultrasound in Chemistry.* Ellis Horwood, Chichester.)

where f_r is the relaxation frequency, A is the relaxation amplitude and B is the high-frequency residual absorption which is frequency independent. If more than one relaxation process (i.e. n processes) can occur, eqn (4.41) is more accurately written as eqn (4.42):

$$\frac{\alpha}{f^2} = \sum_{i=1}^{i=n} \frac{A_i}{1 + (f/f_{ri})^2} + B \qquad (4.42)$$

The loss per cycle, or absorption per unit wavelength, μ, relating to the relaxation is

$$\mu = \alpha'\lambda \qquad (4.43)$$

where α' is the excess absorption for the relaxation process. Thus,

$$\mu = (\alpha - Bf^2)\lambda \qquad (4.44)$$

or

$$\mu = \frac{Acf}{1 + (f/f_r)^2} \qquad (4.45)$$

This curve (Fig. 4.10 curve B) reaches a maximum value ($= \mu_m$) for μ when $f = f_r$, that is when

$$\mu_m = \frac{Acf_r}{2} \qquad (4.46)$$

In qualitative terms, the shape of the velocity-frequency curve may be explained in terms of increasing 'stiffness' (and hence elasticity, E) of the medium, because with increasing frequency the medium finds it more and more difficult to transfer the translational energy into other forms. Since velocity is related to elasticity by eqn (4.47), the velocity will also increase. This change in velocity with frequency is called *velocity dispersion* and may be represented by eqn (4.48), where the subscripts 0 and ∞ refer to the sound velocity, c, at low and high frequencies:

$$c = \left(\frac{E}{\rho}\right)^{1/2} \tag{4.47}$$

$$c^2 - c_0^2 = \frac{(2\mu_m/\pi)c_0c_\infty(f/f_r)^2}{1 + (f/f_r)^2} \tag{4.48}$$

4.4.1 Determination of energy parameters

Figure 4.11 shows the energy-level diagram for a two-state equilibrium (eqn (4.49)), which may be thought of as either structural, conformational or chemical, and in which state B is assumed to be the higher in energy. By determining the temperature dependences of f_r and the attenuation per wavelength (μ), it is possible to obtain information on the rate constants for the forward and back reactions, k_f and k_b, the free energy barrier, $\Delta G_b^{\ddagger}$, and the free energy difference, ΔG.

$$A \underset{k_b}{\overset{k_f}{\rightleftharpoons}} B \tag{4.49}$$

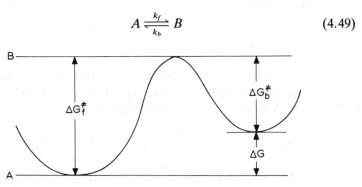

Fig. 4.11 Energy-level diagram for a two-state equilibrium. (Reproduced with permission from Mason, T. J. & Lorimer, J. P. (eds) (1988). *Sonochemistry: Theory, Application and Uses of Ultrasound in Chemistry.* Ellis Horwood, Chichester.)

For example, if we assume for the above equilibrium that each reaction is unimolecular and takes place in an ideal solution, then the characteristic relaxation frequency, f_r, can be related to the rate constants (see below) by eqn (4.50):

$$f_r = \frac{1}{2\pi\tau} = \frac{k_f + k_b}{2\pi} \qquad (4.50)$$

where τ is the relaxation time of the equilibrium. Since for any reaction the rate constant, k_r, can be expressed in terms of the free energy of activation, $\Delta G^{\ddagger}$ by

$$k_r = \frac{kT}{h}\exp\left(-\frac{\Delta G^{\ddagger}}{RT}\right) \qquad (4.51)$$

then eqn (4.50) may be written as

$$f_r = \frac{kT}{2\pi h}\left[\exp\left(-\frac{\Delta G_f^{\ddagger}}{RT}\right) + \exp\left(-\frac{\Delta G_b^{\ddagger}}{RT}\right)\right] \qquad (4.52)$$

If we assume that the equilibrium in the above reaction lies well to the left, then we may assume that $\Delta G_f^{\ddagger} \gg \Delta G_b^{\ddagger}$ such that eqn (4.52) becomes

$$f_r = \frac{kT}{2\pi h}\exp\left(-\frac{\Delta G_b^{\ddagger}}{R\tau}\right) \qquad (4.53)$$

or, since $\Delta G = \Delta H - T\,\Delta S$,

$$f_r = \frac{kT}{2\pi h}\exp\left(\frac{\Delta S_b^{\ddagger}}{R}\right)\exp\left(-\frac{\Delta H_b^{\ddagger}}{RT}\right) \qquad (4.54)$$

Thus, if f_r is measured over a temperature range, $\Delta H_b^{\ddagger}$ and $\Delta S_b^{\ddagger}$ may be determined from the slope and intercept of the $\ln(f_r/T)$ vs $1/T$ plot. To determine ΔH, ΔG and ΔS, use is made of eqn (4.55):

$$\mu_m = \frac{\pi R}{2C_p}(\gamma - 1)\left(\frac{\Delta H}{RT}\right)^2\exp\left(-\frac{\Delta H}{RT}\right)\exp\left(\frac{\Delta S}{R}\right) \qquad (4.55)$$

In practice the values of C_p (molar heat capacity at constant pressure) and γ may not be known with sufficient accuracy, and it is preferable to replace one of them with the measured velocity of sound (eqn (4.56)),

$$c^2 = [(\gamma - 1)C_p]/(T\theta^2) \qquad (4.56)$$

where θ is the expansion coefficient. Substitution into eqn (4.55) yields

$$\frac{\mu_m T}{c^2} = \frac{\pi \theta^2}{2R} \left(\frac{\Delta H}{C_p}\right)^2 \exp\left(-\frac{\Delta H}{RT}\right) \exp\left(\frac{\Delta S}{R}\right) \qquad (4.57)$$

Relaxation in polymers may be divided into processes involving segmental motion of the backbone or side groups (independent of RMM) and overall co-operative motion of the whole backbone (RMM-dependent). Prior to the use of acoustic relaxation techniques, the study of segmental (and side chain) motion was restricted mainly to dielectric relaxation studies on polar molecules (dielectric studies are unable to provide information regarding motion of a non-polar molecule), with investigation of whole-chain movements[37] restricted to a study of the frequency dependence of the shear viscosity of the polymer solution.[38,39]

Most experimental investigations of polymers use frequencies in the range 100 kHz–500 MHz, with the attenuation determined as a function of temperature, frequency, polymer concentration and relative molar mass. Equation (4.42) is fitted reiteratively by computer to the experimental data[40] in the frequency range investigated and the best-fit values of A, f_r and B are obtained by assuming that either single, double or multiple relaxation phenomena are involved.

In general, the dynamic spectrum of a polymer may be divided into two parts. The first is a low-frequency process, with a relaxation time ($\tau_r = 1/(2\pi f_r)$) which is molecular-weight-dependent[41–44] and an amplitude (A) which correlates with the viscosity of the solution. The second is a high-frequency process which is molecular-weight-dependent for low molecular weights and independent for higher molecular weights.[45,46]

Single-relaxation models[47] (i.e. one value of A and f_r) have been interpreted in terms of segmental motion of the backbone, whereas double relaxation models (A_1 and f_{r1}; A_2 and f_{r2}) have been interpreted in terms of motion of the backbone and the side groups. For example, the data for several polyvinyl esters[48–51] has been found to fit a double-relaxation model yielding two values of A and f_r (Table 4.8).

The low-frequency f_r value (3–8 MHz), being almost independent of the length of the side chain, is interpreted as being due to motion of the backbone. The high-frequency f_r value (60–150 MHz) decreases significantly with increase in the length of the side chain and is thought to be associated with reorientational motion of the side chain.

A study of the dependence of the acoustic absorption coefficient (α)

Table 4.8 Relaxation parameters for various polyvinyl esters in toluene

Polymer	Temperature (K)	A_1	A_2	f_1 (MHz)	f_2 (MHz)
		$[(a/f^2) \times 10^{15} s^2 m^{-1}]$			
Poly(vinyl acetate)	273	64·5	6·2	7·9	150
	283	46·5	6·2	8·2	150
Poly(vinyl proprionate)	283	41·7	7·7	6·2	105
	293	32·6	7·7	6·5	106
Poly(vinyl butyrate)	273	30·5	15·7	3·2	57
	283	24·3	15·0	3·4	60
	293	18·2	12·0	4·5	63

on polymer concentration, C, has in some cases yielded breaks in the α vs C curves.[43,45,52,53] These break points have been ascribed[54] to the increased polymer–polymer interactions which occur with the onset of chain entanglement[55] in the solution.

As with pure liquids, studies of the dependence of attenuation on temperature has allowed a determination of the thermodynamic parameters (ΔG, ΔH, ΔV, ΔS) associated with the various conformational changes.[56] For example, the activation energy for polystyrene in CCl_4[45], obtained by plotting the acoustic relaxation time against $1/T$, is in good agreement ($27.3\,kJ\,mol^{-1}$) with the values obtained from dielectric studies[57] in poly(p-chlorostyrene) ($21\,kJ\,mol^{-1}$) and NMR measurements[58] in the same solvent.

The energies associated with the conformational change, however, depend not only on the nature of the group attached to the backbone, but also upon the configuration (i.e. tacticity[55]) of the polymer. Certain configurations will have conformations which require lower activation energies to achieve a particular spatial arrangement than do others. For example, the acoustic energy difference between the conformational states of poly(α-methylstyrene) (PMS), when predominantly syndiotactic, is greater than when the polymer is predominantly isotactic. This PMS value ($8.3\,kJ\,mol^{-1}$) is also greater than that for the less hindered polystyrene chain ($5.4\,kJ\,mol^{-1}$). For poly(methyl methacrylate),[60] the energy differences for syndiotactic, atactic and isotactic are 6·3, 6·3 and $3.7\,kJ\,mol^{-1}$, respectively. Changes in the tacticity of the polymer also appear to have a marked influence on both the position and amplitude of the variation of the absorption coefficient with frequency.

Polyelectrolytes, combining the properties of polymers (chain flexibility) with those of electrolytes (strong electrostatic interaction) have also been investigated using ultrasonic relaxation methods.[46,61] Because of the many processes which, theoretically, could give rise to excess ultrasonic absorption, e.g. segmental motion of backbone and side groups, solvation, proton transfer and ion-pair formation, caution must be exercised in assigning the relaxations to a particular process.

Ultrasonic attenuation measurements have been employed for purposes other than the determination of the energy barriers for conformational change. For example, they have been used in the analytical sense

   (a) to investigate the nature of the dynamic behaviour of macromolecules under the action of an elastic strain;[62]
   (b) to monitor polymer and composite quality;[63]
   (c) to investigate the morphological changes which take place in the suspension polymerization of poly(vinyl chloride);[64]
   (d) to provide information on particle growth during the emulsion polymerization of poly(vinyl acetate).[65]

In the last two applications, measurement of the ultrasonic velocity with reaction time has been shown to be an effective method of continuously monitoring the polymerization process.

4.5 Polymer processing and technology

There are at present only a few commercial applications of ultrasound in the plastics industry. The best known is probably the welding of thermoplastics, a process which now lends itself readily to automation. In common with the welding of metals, the ultrasonic welding of plastics is primarily a hot-stage process, except that it is the alternating high-frequency stress which generates the heat in the plastic causing it to melt. Provided the ultrasonic heat is applied selectively at the interface only, it will cause minimum distortion and degradation of the material. Obviously in order that a good weld may be obtained, the plastic must have characteristics which make it suitable for welding. These include

   (1) the ability to transmit and absorb the vibrational energy;
   (2) a relatively low melting/softening temperature;
   (3) a low thermal conductivity to facilitate local build-up of heat;
   (4) a low proportion of lubricant, plasticizer or trapped moisture, which tend to have adverse effects on weldability.

Crystalline polymers with high melting temperatures and a very narrow melting range are generally difficult to weld ultrasonically, whereas the rigid amorphous plastics (e.g. polycarbonate or polystyrene) are best. In fact, because of their high mechanical Q, the rigid plastics can be 'far-field' welded (i.e. welded at some distance from contact owing to the low energy attenuation), whereas the softer or rubber-modified and porous plastics, because of their high damping are only suitable for 'near field' welding.

A typical ultrasonic welder consists of a high-frequency electrical power supply, an electromechanical transducer and a mechanical, usually pneumatically powered, press for clamping the parts during welding. Although welding times of the order of 1/10 of a second are common, most welds are accomplished in about a second or so. This allows, on completion of the ultrasonic exposure, a short delay before retraction of the horn to enable the plastic time to solidify. Although it is quite possible to use either 10 or 40 kHz instruments, most ultrasonic welders now in use operate at 20 kHz and a power density (i.e. intensity) of $1000 \, W \, cm^{-2}$. The choice of 20 kHz represents a compromise in equipment cost, noise (10 kHz in the audible range) and the ability to handle the particular application (lower frequencies can sustain greater intensities).

With the improvement in equipment design and the manufacture of higher-powered instruments have come the welding-related applications such as staking, metal-in-plastic insertion and spot-welding.

The principle of ultrasonic staking, which is the enclosure of another material in plastic, is shown in Fig. 4.12. Here the ultrasonic stress induced in the plastic causes the stud to yield and eventually conform to the shape of the horn. The advantages of this technique are that

(a) tight joints are obtained;
(b) the lower temperatures minimize the extent of degradation and allow the use of crystalline materials.

Another method of securing metal to plastic is ultrasonic insertion. In this case a hole is pre-moulded in the plastic (slightly smaller than the metal insert) and the horn is used first to melt the plastic before driving the insert into place (Fig. 4.13).

Ultrasonic spot welding is a process which is being used increasingly as an alternative to adhesion, riveting and stapling in the automobile and furniture industries. Using specially designed projection tips,

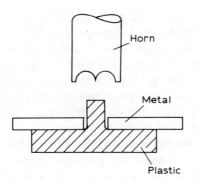

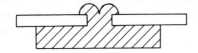

Fig. 4.12 Ultrasonic staking with plastics.

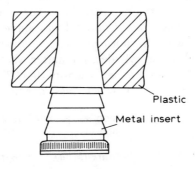

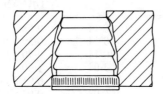

Fig. 4.13 Ultrasonic insertion with plastics.

enhanced weld strength (and good appearance) is produced in such plastics as ABS, polyethene, polypropene and PVC.

Ultrasound has also been used to advantage in the moulding of plastic powders. For example, Fairbanks[66] and separately Paul and Crawford[67] have shown that it is possible to mould both thermoplastic (acrylics and vinyls) and thermoset powders (phenolics and allylics) by the application of high-intensity ultrasound (20 kHz; 150 W and 900 W, respectively) and pressure without the use of external heat. The results of such studies indicated that whereas the effect of pressure was insignificant, the smaller the particles employed ($\sim$150 μm) and the longer the dwell time (up to 6 s) the greater was the tensile strength of the material produced. In fact, Paul and Crawford found that polypropene moulded 'ultrasonically' had 85% of the strength obtainable by conventional injection moulding. Unfortunately there was a trade-off in ductility, with the ultrasonically moulded material having only 15% extension to break as compared with the 'norm' of 300%.

Fairbanks[68] has also studied the effect of ultrasonic energy on the flow characteristics of a poly(methyl methacrylate) melt in a simulated injection moulder. Initially the ultrasound (20 kHz, 0–105 W) was applied either simultaneously or independently to both the extruder tube and the cylinder of the moulder. However, since no discernible effect was observed when the ultrasound was applied at the extruder tube, further work with the horn in this position was discontinued. Applying ultrasound to the cylinder, no matter what the position, led to a significant decrease in the melt viscosity (0–60%), and increase in flow rate, the magnitude of which was dependent on the melt

Table 4.9 Viscosity reduction as a function of ultrasonic intensity

Temperature (°C)	Pressure (MPa)	Power level (W)		
		45	75	105
		$(\eta(US)/\eta(Initial))$		
185	10·14	0·58	0·45	0·39
190	10·14	0·59	0·41	0·35
195	10·14	0·70	0·43	0·39
200	3·04	1·00	0·96	0·82
200	5·07	0·91	0·85	0·69
200	6·08	0·92	0·77	0·57

temperature, the ram pressure and the ultrasonic power level (Table 4.9).

Fairbanks' suggestion was that the reductions were due to one of the following.

(1) Creation of a slip flow at the wall of the cylinder and the entrance to the extruder tube. No doubt this is due to the preferential absorption of ultrasonic energy at the melt–metal interface which then appears as heat, thereby reducing the viscosity at the wall.

(2) Reduction in the apparent bulk viscosity due to a change in polymer rheology. We have already seen in Section 4.2 how ultrasound can lead, via degradation, to a reduction in polymer solution viscosity. Although Fairbrother did not investigate whether degradation of the polymer, and subsequent reduction in RMM and hence viscosity had occurred, it seems reasonable to assume that the polymer melt with an initial viscosity of 30 000–100 000 poise would certainly have resisted cavitation and thus degradation.

One disadvantage of the use of ultrasound in the process was that it proved extremely difficult, unless external heat was supplied, to remove the remaining polymer slug from the cylinder. This seems to indicate that ultrasound had also increased the magnitude of the polymer-to-metal bond.

An application[69] which utilizes both ultrasound's ability to generate heat and a polymer's poor thermal conductivity is in the vulcanization of rubber. Because of rubber's poor thermal conductivity, the conventional process is slow and energy-intensive and typically several hours may be required to vulcanize thick sections of material even at temperatures and pressures of 205°C and 14·3 MPa. Even under these extreme conditions, poor bonding between the non-polar rubber and the various polar reinforcements (e.g. cord, fabric or metal) often results, leading to interfacial failure. Any attempt to counteract these problems, such as the use of various additives to accelerate the cure process, or special surface treatments to improve the bonding, simply serve to increase the manufacturing costs. There is also the further problem that the additives often cause scorching during the moulding stage. However, the use of ultrasound largely overcomes many of these disadvantages. For example, the poor thermal conductivity, a major problem in the conventional process, becomes an advantage in the ultrasonic process. In the conventional process, poor conductivity resists the penetration of the heat, which is applied externally and often leads to charring. In the ultrasonic process, the heat is generated internally and is stopped from escaping. It also allows a more uniform internal build-up of temperature.

Ultrasonic vulcanization also tends to change the interfacial property of the rubber and the reinforcing materials to improve bonding. Improved wetting and flow characteristics produced by ultrasonic vulcanization have the potential to increase the interfacial bond strength between the rubber and the reinforcing materials currently used.

Using frequencies of 10–100 kHz has led to more than 100% increase in the production rate; better than 50% energy saving; potential of improved bond qualities resulting in better mechanical and ageing properties of the final product; reduced raw materials cost resulting from reductions or elimination of costly additives such as accelerators, activators and coupling agents; lower mould temperatures; and accelerated degassing. It does require, however, that the process is performed at an overpressure of 500–1000 p.s.i. to inhibit cavitation and possible degradation.

In conclusion, let us consider the application of ultrasonic technology to polymer blending. Perhaps the major difficulty encountered in blending is the production of a homogeneous mixture with complete dispersion of the added component. Scott-Bader UK, using a whistle reactor, have achieved considerable improvement in the blending of pyrogenic silica with various polyester resins. In that the whistle is capable of processing 12000 litres/h and provides the correct thixotropic characteristics, no matter what resin is employed, emphasizes the possible capabilities of ultrasound in large industrial applications, and in doing so answers the question of whether this type of novel technology possesses a future.

4.6 References

1. G. Stokes, *Trans. Camb. Phil. Soc.*, **8**, 287 (1849).
2. G. Kirchoff, *Ann. Phys. (Liepzig)*, **134**, 177 (1868).
3. E. A. Neppiras, *Phys. Rep.*, **61**, 160 (1980).
4. M. E. Fitzgerald, V. Grffing & J. Sullivan, *J. Chem. Phys.*, **25**, 926 (1956).

5. G. Schmid & O. Rommel, *Z. Phys. Chem.*, **A185**, 97 (1939); *idem.*, *Z. Electrochem.*, **45**, 659 (1939). G. Schmid and E. Beuttenmuller, *Z. Elektrochem.*, **49**, 325 (1943); *idem.*, *Z. Elektrochem.*, **50**, 209 (1944). G. Schmid, *Phys. Z.*, **41**, 326 (1940); *idem. Z. Phys. Chem.*, **186A**, 113 (1940). G. Schmid, P. Paret & H. Pfleider, *Kolloidn. Zh.*, **124**, 150 (1951).
6. H. F. Mark, *J. Acoust. Soc. Am.*, **16**, 183 (1945).
7. S. L. Malhorta, *J. Macromol. Sci. Chem.*, **A18**, 1055 (1982).
8. W. Gaertner, *J. Acoust. Soc. Am.*, **26**, 977 (1954).
9. M. A. K. Mostafa, *J. Polymer Sci.*, **33**, 295 (1958); *ibid.*, **33**, 311 (1958); *ibid.*, **33**, 323 (1958).
10. M. S. Doulah, *J. Appl. Poly. Sci.*, **22**, 1735 (1978).
11. A. Weissler, *J. Appl. Phys.*, **21**, 171 (1950); *J. Chem. Phys.*, **18**, 1513 (1950); *J. Acoust. Soc. Am.*, **23**, 370 (1951).
12. R. O. Prudhomme & P. Graber, *J. Chim. Phys.*, **46**, 667 (1949). R. O. Prudhomme, *J. Chim. Phys.*, **47**, 795 (1950).
13. H. W. Melville & A. J. R. Murray, *J. Chem. Soc. Faraday. Trans.*, **46**, 996 (1950).
14. H. H. Jellinek & G. White, *J. Polymer Sci.*, **6**, 745 (1951); *ibid.*, **6**, 757 (1951); *ibid.*, **7**, 33 (1951); *ibid.*, **13**, 441 (1954).
15. E. Wada & H. Nakane, *J. Sci. Research Inst. (Tokyo)*, **45**, 1 (1951).
16. C. Keqiang, S. Ye, L. Huilin & X. Xi, *J. Macromol. Sci. Chem.*, **A22**, 455 (1985); *ibid.*, **A23**, 1415 (1986).
17. P. A. R. Glynn, B. M. E. Van der Hoff & P. M. Reilly, *J. Macromol. Sci.*, **A6**, 1653 (1972); *ibid.*, **A7**, 1695 (1973).
18. M. Tabata & J. Sohma, *Chem. Phys. Lett.*, **73**, 178 (1980); *Eur. Polym. J.*, **16**, 589 (1980).
19. J. P. Lorimer, T. J. Mason & D. Kershaw, *Ultrasonics International 89, Conference Proceedings*, Butterworth Scientific, London, in press.
20. Ch'ien Jen-Yuan, *Determination of Molecular Weights of High Polymers*, Israel Program for Scientific Translation, 1963.
21. A. Henglein, *Z. Naturforsch., B*, **7**, 484 (1952); *ibid.*, **10**, 616 (1955); *Makromol. Chem.*, **14**, 15 (1954); *ibid.*, **15**, 188 (1955); *ibid.*, **18**, 37 (1956).
22. A. A. Berlin, *Usp. Khim.*, **29**, 1189 (1960); *Khim. Nauka. Prom.*, **2**, 667 (1957).
23. K. F. Driscoll & A. U. Sridhari, *J. Appl. Polym. Sci., Appl. Polym. Symp.*, **26**, 135 (1975).
24. O. Lindstrom & O. Lamm, *J. Phys. Colloid Chem.*, **55**, 1139 (1951).
25. P. Kruus, D. J. Donaldson & M. D. Farrington, *J. Phys. Chem.*, **83**, 3130 (1979).
26. R. C. Weast, ed., *Handbook of Chemistry and Physics*, 57th ed. Chemical Rubber Co., Cleveland, 1976.
27. T. Miyata & F. Nakashio, *J. Chem. Eng. Jpn.*, **8**, 463 (1975).
28. J. P. Lorimer & T. J. Mason, *Ultrasonics International 87, Conference Proceedings*. Butterworths, London, p. 762.
29. Y. Hatate, T. Ikeura, M. Shinonome, K. Kondo & F. Nakashio, *J. Chem. Eng. Jpn.*, **4**, 38 (1981).

30. J. P. Lorimer, T. J. Mason, K. Fiddy, D. Kershaw, D. Dodgson & R. Groves, *Ultrasonics International 89, Conference Proceedings,* Butterworths (in press).
31. L. Toppare, S. Eren & U. Akbulut, *Polymer Commun.,* **28,** 36 (1987). U. Akbulut, L. Toppare & B. Yurttas, *Polymer,* **27,** 803 (1986); *idem. B. Poly. J.,* **18,** 273 (1986).
32. S. Osawa, M. Ito, K. Tanaka & J. Kuwano, *Synthetic Metals,* **18,** 145 (1987).
33. R. A. Pethrick, *J. Macromol. Sci. Rev., Macromol. Chem.,* **9,** 91 (1973).
34. A. M. North & R. A. Pethrick, in *International Reviews of Science, Physical Chemistry Series 1,* ed. A. D. Buckingham & G. Allen. Butterworths, London, 1972.
35. K. F. Herzfeld & T. A. Litontz, *Adsorption and Dispersion of Ultrasonic Waves.* Academic Press, New York, 1959.
36. R. A. Pethrick, *Sci. Prog.,* **58,** 563 (1970).
37. J. D. Ferry, *Viscoelastic Properties of Polymers.* Wiley, New York, 1971.
38. B. H. Zimm, *J. Chem. Phys.,* **24,** 269 (1956).
39. P. E. Rouse, *J. Chem. Phys.,* **21,** 1272 (1953).
40. T. Sano & Y. Yasunga, *J. Phys. Chem.,* **77,** 2031 (1973).
41. M. A. Cochran, A. M. North & R. A. Pethrick, *J. Chem. Soc., Faraday Trans. 11,* **70,** 1274 (1974).
42. H. Hassler & H. J. Bauer, *Kolloidn. Zh.,* **230,** 194 (1969).
43. W. Ludlow, E. Wyn-Jones & J. Rassing, *J. Chem. Phys. Lett.,* **13,** 477 (1972).
44. B. Fruelich, C. Noel & L. Monnerie, *Polymer,* **20,** 529 (1979).
45. H. J. Bauer, H. Hassler & M. Immendorfer, *Faraday Discuss. Chem. Soc.,* **49,** 238 (1970).
46. S. Kato, N. Yamauchi, H. Nomura, & Y. Miyahara, *Macromolecules,* **18,** 1496 (1985).
47. A. Juszkiewicz, A. Janowski, J. Ranachowski, S. Wartewig, P. Hauptmann & L. Alig, *Acta Polymerica,* **36,** 147 (1985).
48. H. Nomura, S. Kato & Y. Miyahara, *J. Mater. Sci. Jpn.,* **21,** 476 (1972).
49. H. Nomura, S. Kato & Y. Miyahara, *J. Chem. Soc. Jpn., (Chem. Ind. Chem.),* 1241 (1972); *ibid.,* 2398 (1973).
50. Y. Masuda, H. Ikeda & M. Ando, *J. Mater. Sci. Jpn.,* **20,** 675 (1971).
51. O. Funschilling, P. Lemarechal & R. Cerf, *Chem. Phys. Lett.,* **12,** 365 (1971).
52. R. Cerf, R. Zana & S. Candau, *C. R. Seances Acad. Sci.,* **252,** 2229 (1961); *ibid.,* **254,** 1061 (1962).
53. P. Row-Chowdhury, *Indian J. Chem.,* **7,** 692 (1969).
54. H. Nomura, S. Kato & Y. Miyahara, *Nippon Kagaku Zasshi,* **88,** 502 (1967); *ibid.,* **89,** 149 (1968); *ibid.,* **90,** 250 (1969).
55. H. R. Berger, G. Heinrich & E. Straube, *Acta Polymerica,* **37,** 226 (1986).
56. S. Nishikawa & R. Shinohara, *J. Solution Chem.,* **15,** 221 (1986).
57. W. H. Stockmayer, H. Yu & J. E. Davis, *Polymer Preprints,* **4,** 132 (1963).
58. D. W. McCall & F. A. Borey, *J. Polym. Sci.,* **45,** 530 (1960).

59. J. H. Dunbar, A. M. North, R. A. Pethrick & D. B. Steinhauer, *J. Chem. Soc., Faraday Trans. II.*, **71**, 1478 (1975).
60. C. Tondre & R. Cerf, *J. Chem. Phys.*, **65**, 1105 (1968).
61. R. Zana, *J. Macromol. Sci. Revs., Macromol. Chem.*, **C12**, 165 (1975).
62. M. A. Sidkey & Abd el Aal, *Acustica*, **60**, 264 (1986).
63. W. N. Reynolds, L. P. Scudder & H. Pressman, *Polymer Testing*, 325 (1986).
64. P. Sladky, I. Pelant & L. Parma, *Ultrasonics*, **16**, 32 (1979); P. Sladky, I. Parma & I. Zdrazil, *Bull. Polymer*, **7**, 401 (1982).
65. P. Hauptmann, F. Dinger & R. Sauberlich, *Polymer*, **26**, 1741 (1985).
66. H. V. Fairbanks, *Ultrasonics*, 22 (1974).
67. D. W. Paul & R. J. Crawford, *Ultrasonics*, 23 (1981).
68. P. Khaladkar, J. Sears & H. Fairbanks, *Proc. M. Va. Acad. Sci.*, **45**, 412 (1973).
69. N. Senapati & D. Mangaraj, *US Patent*, No. 4 548 771 (1985).

5 Equipment

T. J. Goodwin
Harwell Laboratory, Oxon, UK

5.1 Introduction

The recent rapid growth in interest in sonochemistry is due in part to the availability of relatively inexpensive ultrasonic laboratory equipment. This equipment is usually in the form of an ultrasonic cleaning bath or a sonic probe (or horn). It is interesting to note that both types of equipment were originally designed for other applications: ultrasonic baths for the cleaning of dirty laboratory glassware and sonic probes for biological cell disruption. Both have been readily adapted by the research chemist for laboratory work and their use in sonochemistry is discussed in detail below (Section 5.3).

This chapter is intended to provide the reader with a critical review of the various pieces of laboratory equipment that are available for use in sonochemical research. The advantages and disadvantages of each technique are presented and some indication of the cost of the equipment is given. The addresses of commercial suppliers are provided to give the reader easy access to the equipment. Possible methods for the scale-up of sonochemical reactions are also discussed.

The increase in the number of sonochemical publications in the open literature over the past five years has stimulated great interest amongst industrial chemists. This increasing awareness of the benefits of applying ultrasound to chemical reactions has resulted in many industrial chemists considering the possibilities of scaling-up sonochemical reactions. This in turn has led to various designs of 'sonoreactor' being proposed; these are discussed in Section 5.4.

Before the different types of laboratory and large-scale equipment are reviewed, a brief discussion of the various components and terms used in sonochemical equipment is given (Section 5.2).

5.2 Ultrasonic terminology

5.2.1 Equipment

Ultrasonic cleaning baths and probes have the same basic arrangement in that they contain an electrical generator and an ultrasonic transducer (Fig. 5.1). The generator provides high-voltage pulses of energy at the desired frequency to the transducer. The transducer is the most important component in the system and in virtually all pieces of modern ultrasonic equipment operates on the piezoelectric principle. A transducer is defined as 'a device that is actuated by power from one system to supply power to a second system'; piezoelectric transducers in cleaning baths and probes convert electrical energy to ultrasonic energy.[1]

The pre-stressed piezoelectric design used in this equipment consists of a number of piezoelectric elements (typically two or four) bolted between a pair of metal end masses (Figs 5.2 and 5.3). The two piezo elements are pre-polarized and this enables them to be positioned (electrically opposed but mechanically aiding) such that the end masses are at an earth potential. The high-tensile bolt clamps the

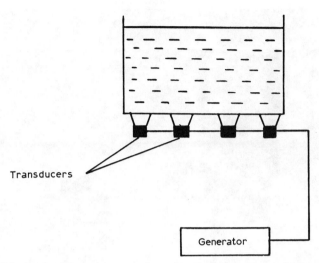

Fig. 5.1 Basic ultrasonic system incorporating generator and transducers.

complete assembly together and keeps the piezo elements under compression. Piezoelectric transducers constructed in this manner have potential efficiencies of 98% for converting the electrical energy to acoustic energy. They can handle power transfers of up to 1000 W when operated on a continuous basis.

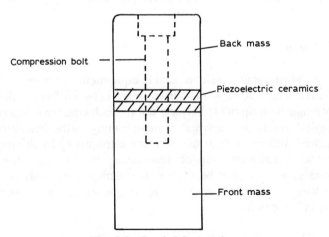

Fig. 5.2 Sandwich transducer.

Fig. 5.3   Piezoelectric transducers: (a) 20 kHz; (b) 40 kHz. (Courtesy of Sonic
Systems)

5.2.2   Ultrasound

Frequency   Most commercially available equipment operates at either
20 or 35 kHz. There is no scope for scanning between frequencies (as
in photochemical equipment) because the piezoelectric transducers are
fixed-(single-) frequency devices. Some cleaning baths irradiate the
cleaning fluid with more than one ultrasonic frequency by driving the
piezoelectric transducers out of resonance. Whilst the range of
frequencies generated may be of use to cleaning operations there is
little evidence to suggest that this technique is of any benefit to
sonochemical research.

Intensity   The maximum ultrasonic intensity applied to a reaction mixture is generally defined in terms of the power density at the radiating face of the ultrasonic transducer. Power density is defined as the electrical power into the transducer divided by the surface area of the transducer (or the end area of the probe tip).

Cleaning baths are low-intensity ($<10\,\mathrm{W\,cm^{-2}}$) systems, whereas probe systems are regarded as high-intensity ($>100\,\mathrm{W\,cm^{-2}}$) systems.

Amplitude The amplitude is taken as the peak-to-peak displacement of the transducer radiating face; at 20 kHz this amplitude is of the order of 1–20 μm.

5.3 Laboratory equipment

There are two basic methods for the introduction of ultrasonic radiation to a reaction system on the laboratory scale. The reaction vessel can be immersed in a liquid (usually water) which is under ultrasonic irradiation or an ultrasonic probe can be introduced directly into the reaction solution. Whichever method is chosen it is essential that both reaction temperature and irradiation power are closely controlled in order to obtain reproducible results. The four methods most commonly described in the literature for laboratory sonochemical research are discussed in detail in the following subsections.

5.3.1 Ultrasonic cleaning baths

This is the most accessible method because of the ready availability of commercial equipment. Numerous companies supply cleaning baths in a range of sizes, usually between 0·5 litres and 100 litres.[2] Most laboratory research is performed in baths of around 10 litres capacity or less. The ultrasound in a cleaning bath system is produced by the transducers mounted on the external walls of the bath; these usually operate at a power density of 1–5 $\mathrm{W\,cm^{-2}}$. The ultrasound is transmitted through the fluid medium, usually water, and into the reaction solution contained within the flask (Fig. 5.4).

The major drawback with this system is that, although there is cavitation in the water outside the flask, because the ultrasound is attenuated at the glass/water interface there is a much reduced intensity ($<0·5\,\mathrm{W\,cm^{-2}}$) within the flask and cavitation may not be present. Three additional drawbacks of this system are as follows.

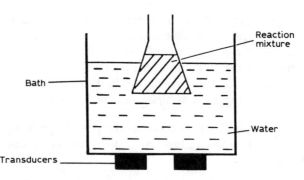

Fig. 5.4 Ultrasonic cleaning bath (transducers mounted on external walls
produce a non-uniform field within the vessel).

*(a) The non-uniform ultrasonic field produced within the
bath* Cleaning baths are usually designed to produce a standing-wave
pattern within the cleaning liquid, maximum ultrasonic intensity being
generated at the antinodes in the pattern. This pattern can be seen by
visual inspection of the surface of the cleaning fluid; such inspection
also reveals that the pattern shifts across the surface with time and is
therefore not a true standing-wave pattern.

Pugin has shown that the ultrasonic intensity within a bath varies
with distance from the transducers.[3] For baths fitted with a single
transducer, the maximum ultrasonic intensity is found above the
transducer; for baths equipped with two transducers the maximum
intensity is found midway between the two (Fig. 5.5). The spatial
distribution of the ultrasonic intensity is significantly affected by a
variety of factors, including depth of water in the bath, operating
voltage of the transducer(s), and shape and position of the flask in the
bath.

The susceptibility of the ultrasonic field within the bath to these
various external factors leads to considerable difficulties in quantifying
the amount of power dissipated, and the cavitation generated, within a
reaction vessel placed in the bath. This also leads to problems with
reproducibility; reaction vessels of similar shape need to be placed in
the same position in the bath for each experiment if reproducible and
consistent results are to be obtained.

(b) Temperature control Cleaning baths warm up during use, particularly over an extended period of time. This is not a problem if the

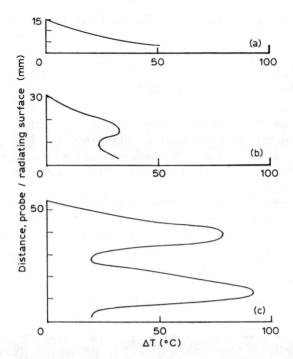

Fig. 5.5 Intensity profiles measured in the Laborette 17 ultrasonic cleaner at various water levels: (a) 15 mm; (b) 31 mm; (c) 54 mm. All profiles were measured in the central axis of the cleaner. (Reproduced with permission from B. Pugin, *Ultrasonics* **25** (1987) 49.)

bath is externally heated but leads to inconsistent results when working at room temperature or below. It is important to continuously monitor the temperature within the reaction vessel as this is usually a few degrees above that of the bath.

(c) Operating frequency and power Although most cleaning baths operate at either 20 or 35 kHz, their operating power varies considerably from one manufacturer to another. Virtually all cleaning baths are fitted only with an on/off switch and a timer to control the period of irradiation. When attempting to repeat literature experiments it is essential to perform the work in the same make of bath as that reported in the paper.

Despite these drawbacks, cleaning baths remain popular amongst sonochemists because they are cheap (cost depends on size—at the time of writing a 5-litre unit can be purchased for around £300) and reliable.

5.3.2 Direct-immersion sonic probes

Sonic probes are rapidly becoming the preferred method of introducing ultrasound to a reaction mixture. They are generally more expensive than cleaning baths (basic models cost around £2000) but their operational advantages (see below) far outweigh this additional expenditure.[4]

Sonic probes consist of the electrically driven transducer mechanically coupled to a probe (or horn) of the desired material (Fig. 5.6). The vibratory motion generated by the transducer is normally too low $(1–10\,\mu m)$ for practical use; the probe acts as an amplifier and generates far greater peak-to-peak displacements at the probe tip $(10–50\,\mu m)$. The probe is 'dipped' directly into the reaction mixture, where it is capable of generating ultrasonic intensities in excess of

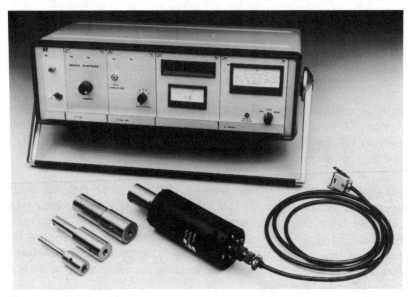

Fig. 5.6 Commercially available sonic probe system comprising generator, probe, and probe tips. (Courtesy of Sonic Systems.)

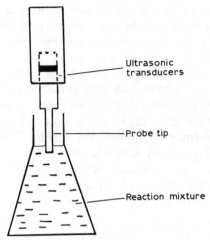

Fig. 5.7 Sonic probe 'dipped' into reaction mixture.

$100 \ W \ cm^{-2}$ (Fig. 5.7). This intensity is far greater than that generated by the cleaning baths and is of considerable use to those sonochemists looking to initiate the formation of reactive intermediates via cavitation of the reaction mixture.

In addition to the high ultrasonic intensities generated by probe systems, other advantages of this technique include the following.

(a) Reproducible operating conditions: advanced probe systems are fully instrumented to monitor transducer amplitude and electrical power delivered to the transducer.[4a,b] This enables the operator to perform sonochemical experiments under controlled ultrasonic conditions.

(b) The direct contact between the probe and the reaction mixtures leads to an efficient transfer (coupling) of ultrasonic energy from the probe to the mixture.

(c) Careful choice of an appropriate probe tip enables a wide range of sample sizes and vessels to be employed in the sonochemical research.

The general disadvantages of sonic probes are as follows.

(a) The zone of high ultrasonic intensity at the probe tip is relatively small; for an end diameter of 5 mm the zone is approximately 7 mm wide and 100 mm long. Thus, reaction mixtures must be vigorously stirred to ensure that all of the reagents experience the high-intensity ultrasound. (Note: in some cases the high-intensity sound provides sufficient agitation to eliminate the need for stirring.)

(b) The high ultrasonic intensities at the probe tip can generate unwanted radicals (via sonolysis of the solvent for example) and these could produce side reactions in the system.

(c) Prolonged use generally results in tip erosion and this may lead to contamination of the reaction mixture. Some companies are now offering probes with interchangeable tip ends (or studs); when one piece becomes severely eroded it can easily be replaced, thus saving on equipment costs and preventing the bulk of the probe becoming eroded.

Because sonic probes now dominate sonochemical research a wide range of accessories and modified probes have become available. The main types of accessory are detailed below.

5.3.2.1 Probe tips Probe tips are usually constructed from titanium alloy, although aluminium, aluminium bronze and stainless steel can be used when required. Titanium alloy is the preferred material of construction because it has high dynamic fatigue strength, low acoustic loss, good resistance to cavitation erosion and, perhaps most importantly for sonochemists, chemical inertness. The probe tip should be half a wavelength long (12·5 cm at 20 kHz) and the intensity generated at the tip (at a given electrical power to the transducer) is inversely proportional to the end area of the tip. A variety of shapes are available; the most common are given in Fig. 5.8 and are described below.

(a) *Linear taper*—the physical shape of this design limits the potential magnification of these probes to approximately fourfold.

(b) *Exponential taper*—this design offers higher magnification than the linear taper (intensities in excess of $150 \, W \, cm^{-2}$ are possible). Its shape is much more difficult to manufacture but the narrow length and small end area makes them particularly suitable for the irradiation of small samples. A number of companies market these exponential tips under the term 'microtip' probes; the end diameter of these microtips is typically 2 mm.[4]

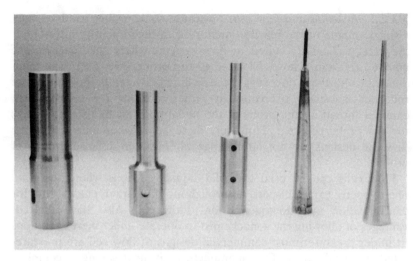

Fig. 5.8 Various shapes of commercially available probe tip. (Courtesy of Sonic Systems.)

(c) *Stepped*—unlike the preceeding designs, the magnification factor is given by the ratio of the end areas. The potential magnification is limited by the tensile strength of the material because considerable stress is generated at the 'step' in the probe. It is an easy design to manufacture and magnifications of 16-fold are readily obtained.

Stepped probes are sold as High-Gain 'Q' horns by equipment suppliers; the end diameter of these probes is usually 25 mm and they are claimed to be particularly suitable for high-intensity processing of larger volumes. The high intensities claimed for such systems are usually between 25 and 50 W cm^{-2}; this is significantly less than those obtained for exponential probes because the end areas of the stepped probes are much greater. The main advantage of stepped probes is their ability to deliver considerably more acoustic power, albeit at a lower intensity, to a reaction solution than can an exponential probe.

(d) *Flow through*—these probes contain an axial bore. The reagents are pumped through the bore and out of the tip directly into the zone of highest ultrasonic intensity. This type of probe is particularly suited to liquid–liquid reactions where emulsification of the two liquids is of prime importance.

5.3.2.2   Reaction and flow cells   Because sonic probes irradiate only
a small volume of the reaction mixture with high-intensity ultrasound,
they are particularly suited to flow systems where the reagents are
pumped through a flow cell fitted to the probe (Fig. 5.9). The main
difficulty with these flow cells is the sealing between the probe and the
cell; this is usually overcome by either screwing the cell onto an
external thread on the probe at the nodal point or by using 'O'-rings
for the probe/cell seal. However, these approaches means that the
flow cell designs are not interchangeable between different makes of
probe.

Flow cells can be used for solid–liquid reactions where the solid
phase needs to be dispersed throughout the liquid phase, and for
emulsification and homogenization. Flow cells also have the advantages of allowing the sonochemist to operate under inert conditions
or under pressure; most commercial designs of flow cell are pressure-
rated up to 100 p.s.i. One manufacturer is now selling a complete
'Sonoreactor' unit which consists of a sonic probe and the appropriate
flow cell incorporated into a loop system with a pump and holding
vessel (Fig. 5.10).[4e]

A particular design of reaction cell is the Suslick Cell and Collar
(Fig. 5.11). These were developed by K. S. Suslick during his

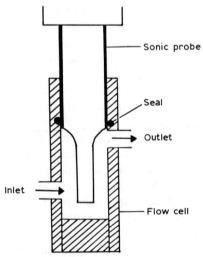

Fig. 5.9 Flow cell fitted to sonic probe.

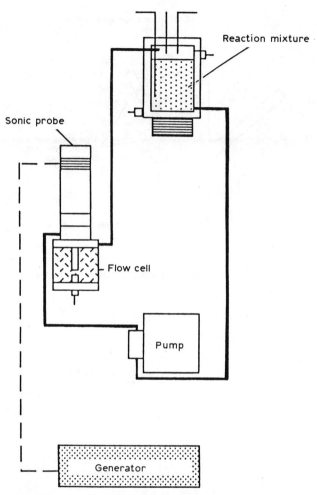

Fig. 5.10 Circuit employed in commercially available 'Sonoreactor'.

sonochemical research into organometallic reactions.[5] A mating collar, complete with gas-tight O-rings, seals the cell to the probe; the cell itself is made of borosilicate glass and has three side arms for introduction and removal of sample and gas and for temperature monitoring.

5.3.2.3 Cooling cells In order to control the temperature of the

T. J. Goodwin

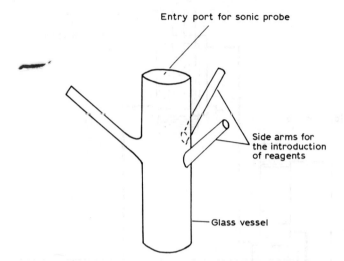

Fig. 5.11 Reaction cell designed by K. S. Suslick.

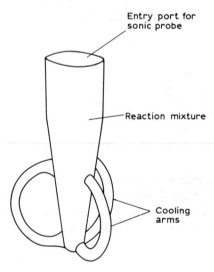

Fig. 5.12 Rosett cell for sonochemical research.

irradiated reaction mixture, some form of external cooling is required. Manufacturers now supply either glass or metal cooling cells which will provide sufficient cooling for the reaction. They usually consist of a sonication vessel contained within a water jacket; the jacket has entry and exit ports for the cooling water. A particular design of cooling cell is the Rosett Cell (Fig. 5.12); the cell is placed in a water bath and the radiative pressure from the probe forces the reaction fluid to circulate through the three cooling arms.

5.3.3 Cup horns

Cup horns can generally be considered as high-intensity water baths. A small plastic jacketed vessel ('cup') is screwed onto a large diameter (>50 mm) probe and the reaction vessel is placed in the water-filled cup (Fig. 5.13). The ultrasound is transmitted from the probe via the water into the reaction mixture; the intensity within the mixture is lower than that in the water and much less than that obtained by using a direct immersion probe. The additional plastic vessel prevents direct intrusion of the reaction mixture by the probe, thus reducing contamination of the mixture and allowing sealed samples to be irradiated. Cooling of the plastic vessel is achieved by flowing water through the external jacket of the vessel.[4]

5.3.4 'Whistle' reactors

Ultrasonic 'whistle' reactors have only been used in a few sonochemical reactions; the most successful application being the enhancement of ester hydrolyses reported by Davidson.[6] The whistle reactor consists of a pump which forces the reaction liquid through the whistle, which contains the ultrasonically vibrating blade. The blade generates cavitation within the mixture (Fig. 5.14). The whistle reactor is particularly suited for liquid–liquid reactions, where the emulsifying effects of the blade will have the greatest effect. They are less suited for solid–liquid reactions because of the difficulties of passing solid material through the whistle. The other major drawback with this method is the lifetime of the stainless-steel blade, which is inevitably reduced when the reaction mixture contains highly corrosive materials.

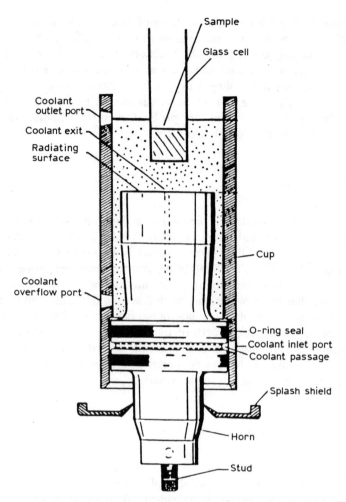

Fig. 5.13 Cup horn for sonochemical research.

5.3.5 *Methods for detecting cavitation*

It is now generally accepted that cavitation is primarily responsible for the ultrasonic enhancement of many chemical reactions. In order to perform reproducible and accurate sonochemical research it is essential that the regions and extent of cavitation within the irradiated reaction mixture are known. This is particularly true for cleaning-bath systems where the ultrasonic intensity, and therefore the extent of

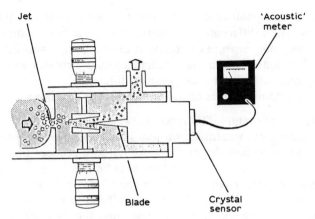

Fig. 5.14 Commercially available 'whistle reactor'. (Reproduced with permission from R. S. Davidson, A. Safdar & B. Robinson, *Ultrasonics* **25** (1987) 35.)

cavitation, varies throughout the bath. There are three empirical methods for detecting the presence and extent of cavitation in the reaction system and these are described below.

5.3.5.1 Erosion of aluminium foil Aluminium foil is easily perforated by cavitation. A small square of foil is 'dipped' in the irradiated fluid for a fixed period of time (typically a few seconds). This results in significant perforation of the foil and the production of erosion patterns on the metal surface. The weight loss due to the perforation reflects the extent of cavitation at that point in the irradiated fluid and the erosion pattern gives an indication of the topology of the cavitation zone. The major drawbacks of this method are as follows.

(a) The presence of the aluminium foil alters the ultrasonic field within the irradiated vessel.
(b) The aluminium is eroded almost too rapidly by high-intensity fields.

5.3.5.2 Thermocouple probe A thermocouple or thermistor coated with a sound-absorbing material may be used to provide a qualitative estimation of ultrasonic intensity within an irradiated vessel.[7] The absorbant material transforms the ultrasonic energy into heat, which is detected by the thermocouple. By measuring the difference between the temperature of the surrounding liquid and the temperature of the thermocouple probe (at equilibrium) an estimate of the ultrasonic intensity at that point in the irradiated vessel can be made; the bigger the temperature difference, the more intense the field. Although this method does not measure cavitation it can be safely assumed that the extent of cavitation within the irradiated liquid increases with increasing ultrasonic intensity (or temperature difference).

The main disadvantages of this technique are as follows.

(a) The temperature difference between the probe and liquid is affected by streaming of the liquid around the probe; the streaming cools the probe and therefore reduces the temperature difference. This streaming may be due to gross mechanical mixing, convective currents, or the ultrasound itself.

(b) The absorbant material is usually silicon rubber and this swells in most organic solvents, reducing the 'sensitivity' of the technique. Cork can be used as an alternative, but the observed temperature differences are much less than those with silicon rubber.

(c) Prolonged exposure to cavitation causes considerable damage to the coating of the probe; replacing this material usually leads to problems with reproducibility, since it is not possible to prepare identical coatings for the thermocouple.

5.3.5.3 Chemical dosimetry This involves the use of a chemical reaction to indicate the presence and extent of cavitation in an irradiated fluid. There are two approaches to this technique.

(a) Irradiation of the whole reaction mixture, with the product being produced only in those areas of highest cavitation activity. A good example is the sonication of a water–carbon tetrachloride mixture, which yields chlorine.[8] The concentration of chlorine produced at any one point in the mixture is dependent on the intensity of the field at that point. Measurement of the local chlorine concentration at each point gives a map of ultrasonic intensity throughout the vessel. It was found that chlorine yields varied from zero to a maximum at distances of $(\frac{\lambda}{4} + n\frac{\lambda}{2})$ from the transducer, where λ is the wavelength of the sound in the medium and n is an integer.

(b) Placing a solid reagent at various points in the ultrasonic field and measuring the rate of the sonochemical reaction between the solid and another reagent dissolved in the reaction liquid. This technique has been reported by Pugin, who used the preparation of organolithium species as the indicating reaction.[3] In this case, as with the aluminium-foil method described above, it must be remembered that placing a solid object in the ultrasonic field alters the spatial distribution of the field in the vessel.

Both approaches assume that cavitation is primarily responsible for the chemical 'indicator' reaction. This is probably true for homogeneous systems (a) but unlikely for heterogeneous systems (b) where the mass-transfer effects of high-intensity ultrasound make a substantial contribution to the enhancement of the indicator reaction.

5.4 Large-scale sonochemical equipment

The successful development of sonochemistry as a general technique for the chemical industry is dependent, in part, on the availability of suitable equipment. There is a need for sonoreactors which will allow sonochemical reactions to be performed on a multi-litre scale under controllable and reproducible ultrasonic conditions. Although this equipment is not available at present, there is considerable interest amongst the UK chemical industry in scaling up sonochemical reactions, which should lead eventually to the required equipment being developed.

This section is therefore intended to inform the reader of the various designs of sonoreactor that have been proposed in the literature. It is divided into two subsections: one deals with the possible designs of sonoreactor for applying ultrasound to large-scale batch reactions, the other is concerned with sonoreactors which may be used to apply ultrasound to continuous reactions. In each case, the advantages and disadvantages of each design are presented and some indication of the probable cost is given.

5.4.1 Batch reactors

Batch reactions are used throughout the chemical industry for a large number of different processes; the reactors used for these processes vary in size but are typically between 3 and 8 m^3. They are, therefore, between 30 000 and 80 000 times bigger than the standard 100-ml flasks normally employed in laboratory sonochemical research. In spite of this large difference in size, the laboratory equipment, such as cleaning baths and sonic probes, used for this research serve as good models for possible commercial-scale sonoreactors. Thus there are three basic designs for sonochemical reactors operating on a batch basis.

5.4.1.1 'Cleaning bath' reactors Large (multi-litre) ultrasonic cleaning baths are readily available from a number of suppliers.[9] They have a large number of transducers fitted to the external walls of the tank and contain the electronics needed to operate the transducers for prolonged periods of time. Thus, the equipment and techniques required to irradiate large volumes of liquid with relatively low intensity ($<10 \, \mathrm{W \, cm^{-2}}$) ultrasound are available at present. This technology could be extended to produce 'cleaning bath' sonoreactors with the transducers mounted on the external walls of a chemical reactor (Fig. 5.15).

The main advantages of this approach to scale-up are that there is no contamination of the reagents by erosion of the ultrasonic source and that the vessel may be operated under pressure without the need for elaborate sealing around the transducers. There are, however,

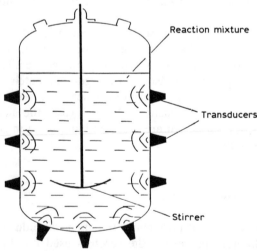

Fig. 5.15 'Cleaning bath' sonoreactor employing wall-mounted transducers.

considerable disadvantages with the design. These include the following.

(a) The transducers have to be permanently mounted to the sides of the vessel. They have to be mechanically fixed in place (usually by adhesives) to achieve efficient transfer of the ultrasound into the reaction mixture. Most existing designs of chemical reactor have curved sides and this in itself poses severe engineering problems because the transducers are flat-faced and are best suited to mounting on flat surfaces. Fixing these transducers to curved faces results in poor adhesion and low energy transfer through the interface. Modern chemical reactors are usually constructed with an outer layer of insulating material or a water jacket; these would have to be modified or removed to allow the transducers access to the inner wall.

(b) The externally mounted transducers generate a low-intensity, non-uniform field within the vessel which will be localized in the vicinity of the reactor walls. Thus, the reagents would have to be stirred vigorously to ensure that all components of the reaction mixture experienced the ultrasound.

(c) The large number of transducers required for this type of reactor would lead to significant bulk heating of the reaction mixture; this is not always desirable.

The problems of mounting and access mean that a new design of chemical reactor would have to be produced or, alternatively, an existing design severely modified in order to accommodate the transducers. This would inevitably be expensive as it would involve the manufacturer in the purchase of new capital equipment; they would much prefer a 'retro-fit' situation where their existing equipment could be easily and cheaply modified.

The high cost of this approach is illustrated by the following example. The number of transducers required for a 5 m^3 cleaning bath sonoreactor may be calculated by considering the surface areas of the reactor and an ultrasonic transducer. The reactor has an external surface area (sides and bottom) of approximately 16 m^2; the surface area of a typical 20 kHz cleaning bath transducer is 3×10^{-3} m^2 (diameter = 6 cm). A 10% coverage of the reactor walls with transducers would require 570 transducers. The volume of reaction mixture irradiated per transducer at this 10% coverage is 9 litres, which is reasonable compared to values of around 6 litres for large cleaning baths. The cost of these transducers would be around £150 000 (at current prices).

Each transducer requires 50 W to produce an intensity of 1–2 W cm^{-2} on the radiating face; the power required to drive these transducers is of the order of 30 kW. This would require 20 generators at a cost of £50 000 (at current prices). Increasing the intensity at the transducer face would require more expensive generators and the overall costs would increase accordingly. Thus, the cost of equipment (excluding the cost of modification and fitting) for a 5 m^3 cleaning bath sonoreactor would be around £200 000.

5.4.1.2 Sonoreactors employing immersible transducers Large immersible transducer arrays have been developed by the manufacturers of ultrasonic cleaning baths as an alternative approach for the cleaning of various components.[10] The immersible units are placed within the cleaning tank and this removes the need for any permanent attachment of the transducers to the vessel walls. Immersible transducers also offer the user considerable flexibility in deciding on the number of arrays he needs to fulfil each cleaning operation. Immersible transducers could be adapted for use in large-scale chemical processing where the required number of arrays are placed within the reaction vessel (Fig. 5.16).

The number of immersible arrays for a 5 m^3 reactor can be calculated on the basis of the number of transducers required and the volume of each array. The typical dimensions of a commercially available immersible transducer are 0·7 m × 0·25 m × 0·1 m; these give a volume of 0·018 m^3 and a surface area (of the radiating face) of 0·18 m^2. This size of array usually contains 32 transducers. Thus, to get 570 transducers (as for the cleaning-bath reactor described in Section 5.4.1.1) into the vessel, you would require 18 arrays. These would occupy a volume of 0·3 m^3, approximately 1/15 of the reactor volume, and have a radiating area of 3 m^2 (1/5 of the vessel wall area). The cost of these arrays would be comparable to that of the ultrasonic equipment required for the cleaning-bath reactor discussed earlier.

The advantages of using immersible transducers for large scale sonochemistry include the following.

(a) Higher ultrasonic intensities would be obtained within the vessel than those achieved inside cleaning bath reactors.

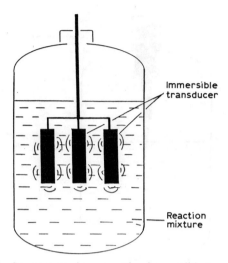

Fig. 5.16 Sonoreactor incorporating immersible transducers.

(b) They offer a retro-fit situation to the chemical manufacturer since, given a suitable number of entry ports, they could be operated without any permanent modification to the reaction vessel.

There are, however, a number of disadvantages associated with this approach.

(i) Erosion of the metallic transducer cover. This can be severe and some manufacturers are now producing chrome-plated covers to reduce the extent of this erosion.
(ii) They produce a non-uniform field within the vessel.
(iii) They are intrusive and will interfere with other mechanical components, such as stirrers, within the reactor vessel.
(iv) Sealing of the transducers and the electrical cables from the reagents. This could pose severe safety problems since most of the solvents used in chemical manufacture are either corrosive or highly inflammable.

5.4.1.3 Reactors employing sonic probes Most laboratory-scale sono-chemical reactions are performed using a sonic probe; the simplest idea for scaling up the reaction is to use a bigger probe in a bigger vessel. Large sonic probes, up to 10 cm in diameter and using 2·5 kW

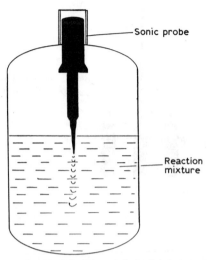

Fig. 5.17   Sonoreactor employing sonic probe for reagent irradiation.

of electrical power, are available commercially.[4b] These units are claimed to be able to handle hundreds of litres per hour on a continuous basis. Their application to batch processing is usually considered to be a question of simply 'dipping' the probe(s) into the reaction mixture (Fig. 5.17).

This approach has two main advantages in that (1) it offers the manufacturers a retro-fit situation and (2) produces very high-intensity fields within the mixture. It has, however, numerous drawbacks.

(a) Only a small proportion of the reaction mixture is irradiated with the high-intensity ultrasound. This is well illustrated by the following calculation. The largest available probe has a 10-cm end diameter and is rated at 2.5 kW (electrical); if all this electrical energy is converted into acoustic power then the intensity at the probe tip is of the order of 120 W cm^{-2}. The volume irradiated with this intensity of ultrasound is approximately 0·001 m^3, which for a 5 m^3 reactor is 1/5000 of the total volume. To irradiate a significant proportion of the reactor volume with this intensity of ultrasound at any one time would, therefore, need a multiple array of probes (at least 10). The cost of such an array at current prices would be in excess of £100 000.

(b) *Erosion of the probe tip.* As for laboratory work, these large sonic probes also erode during use.

(c) *Shielding of the transducer and associated electrical supply from the reaction mixture.* As for the immersible transducers, the probes are fitted within the reaction vessel and this presents severe safety problems to the operators.

(d) *'Stalling' of the probes in high-viscosity fluids.* Sonic probes are unable to operate in high-viscosity reaction mixtures because the viscosity prevents the probe from vibrating at the desired frequency.

5.4.2 Continuous reactors

For the purposes of this chapter, a continuous reactor is considered to be a vessel through which flow is continuous and configured so that process variables are functions of position within the reactor rather than of time. A continuous process can therefore involve either passing the reagents through the reactor once or recirculating the reagents through the reactor.

The main consideration with regards to continuous sonochemical reactions, and one for which there is very little information available at present, is the residence time of the reagents in the ultrasonic field. Is it better to have one slow 'pass' (long residence time) or many fast passes (short residence time) of the reagents through the sonoreactor? Other important considerations for these processes are the dimensions of the irradiated zone and the nature of the reagents in the mixture. For example, if the zone has a 10-litre volume, you will require 500 passes of material to process 5000 litres (assuming 100% conversion per pass). If the conversion is only 10% then 5000 passes are required. Similarly, how are magnesium turnings pumped through a sonoreactor?

There are two main approaches to scaling up continuous sonochemical processes and these are described below.

5.4.2.1 Flow systems incorporating wall-mounted transducers
This approach involves mounting the ultrasonic transducers on the outside of a pipe and irradiating the reagents as they flow through the pipe (Fig. 5.18). This approach has the following advantages.

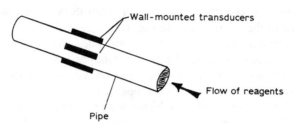

Fig. 5.18 'Flow-through' sonoreactor incorporating wall-mounted transducers.

(i) It is non-intrusive and non-invasive. Contamination of the reaction mixture through erosion of the ultrasonic transducers is unlikely.
(ii) It requires a small number of transducers (fewer than 50).
(iii) Because of the small number of transducers involved it is considerably cheaper than the cleaning bath reactor described earlier.
(iv) It offers the manufacturer a retro-fit option for existing capital equipment.

The disadvantages are the difficulties in mounting the flat transducers on the outside of circular pipes and the low intensities generated within the reaction mixture unless the dimensions of the pipe are correct.

5.4.2.2 Flow systems employing sonic probes Flow systems incorporating a sonic probe fitted into a flow cell are currently available from suppliers who have developed them for the biotechnology industry, where they are used to process large volumes of material up

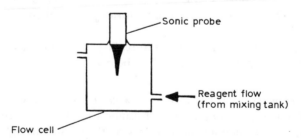

Fig. 5.19 Flow cell and sonic probe for sonochemical processing.

to 1000 litres/h. The probe is mounted on a flow cell and the material is irradiated as it passes through the cell (Fig. 5.19). The main advantages are as follows.

(i) The probes generate high-intensity fields within the cell.
(ii) The flow cell can be fitted in series in the process stream.
(iii) The flow cell and probe can be retro-fitted to existing process streams.

The disadvantages include the following.

(a) Sealing of the reactor/probe joint may present problems particularly for high-pressure reactions.
(b) Erosion of both the probe tip and flow cell.

Both of the above methods for applying ultrasound to reaction mixtures under a continuous or flow regime have one significant advantage over the batch sonoreactors discussed earlier. Because the reaction mixture is pumped through the sonoreactor; the volume of reaction mixture that has to be irradiated at any one time is considerably smaller than if the reaction is processed on a batch basis. Thus, the active zone of a continuous sonoreactor is much smaller in volume and this reduces costs, since fewer transducers and generators are needed to provide the required ultrasonic conditions. Control of the ultrasonic conditions is also easier when trying to irradiate small volumes of fluid. It is therefore preferable to take the reaction mixture to the ultrasound, as for the continuous or flow through sonoreactors, rather than irradiate large volumes of material as in the case of the batch sonoreactors.

5.5 Conclusions

Although there is a range of equipment available for use in laboratory sonochemical research, all the current techniques possess drawbacks which lead to problems with the reproducibility and control of experiments. As the chemical and physical phenomena underlying sonochemical reactions become better understood, so the demand for purpose-built sonochemical equipment will increase. It will become essential that a laboratory sonoreactor which operates under standard and controlled ultrasonic conditions is developed. This will then enable sonochemists to make meaningful comparisons between different reactions and allow the efficiencies and potential of each sonochemical reaction to be assessed.

Similarly, as the full potential of the benefits of applying ultrasound to chemical reactions becomes apparent to industrial chemists, they will require larger-scale equipment which will allow them to operate sonochemical reactions on a production basis. This equipment is not available at present; much work is needed on both the factors controlling scale-up and the ultrasonic equipment itself before production-scale sonoreactors become a reality. An initial assessment of the various designs of commercial sonoreactor indicates that the application of ultrasound to large volumes of reaction mixture, as in the proposed cleaning-bath reactor, is unattractive for a variety of technical and economic reasons. It is much more likely that the eventual scale-up of sonochemical reactions will come through smaller sonoreactors operating on a continuous basis. At the current rate of development, such units should be available within five years.

Acknowledgement

The author gratefully acknowledges the help and advice of Lawrie Ward and John Perkins during the preparation of this manuscript.

5.6 References

1. D. Ensminger, *Ultrasonics,* Marcel-Dekker, New York, 1973.
2. (a) Sonic Systems, Unit 3, Monks Dairy, Isle Brewers, Taunton, Somerset TA3 6QL, UK. (b) Kerry Ultrasonics, Hunting Gate, Wilbury Way, Hitchin, Herts SG4 0TQ, UK. (c) Lucas Dawe Ultrasonics, Concord Road, Western Avenue, London W3 0SD, UK. (d) Maysonic Ultrasonics, Unit 10, Lonlas Village Works, Skewen, West Glam. SA10 6RP, UK. (e) Decon Laboratories, Conway Street, Hove, Sussex BN3 2ZZ, UK. (f) Buehler UK, P.O. Box 150, Binns Close, Coventry CV4 9XJ, UK (g) Jencons Scientific, Cherrycourt Way Industrial Estate, Stanbridge Road, Leighton Buzzard, Beds LU7 8BR, UK. (h) BDH, P.O. Box 8 Dagenham, Essex, RM8 1RY, UK. (i) Langford Electronics, 262–263 Birmingham Factory Centre, Kings Norton, Birmingham B30 3HY, UK.
3. B. Pugin, *Ultrasonics,* **25,** 49 (1987).
4. (a) Sonic Systems, Unit 3, Monks Dairy, Isle Brewers, Taunton, Somerset TA3 6QL, UK. (b) Life Science Laboratories, Sedgewick Road, Luton, Beds, LU4 9DT, UK. (c) Lucas Dawe Ultrasonics, Concord Road, Western Avenue, London, W3 0SD, UK. (d) Roth Scientific, Alpha

House, 9–11 Alexandra House, Farnborough, Hants GU14 6BU, UK. (e) Orsynetics, 2, Venture Road, Chilworth Research Centre, Southampton, Hants SO1 7NP, UK. (f) MSE Scientific Instruments, Sussex Manor Park, Crawley, West Sussex RH10 2QQ, UK.

5. K. S. Suslick, *High Energy Processes in Organometallic Chemistry*, ACS Symposium Series No. 333, 1987, p. 191.

6. R. S. Davidson, A. Safdar & B. Robinson, *Ultrasonics*, **25**, 35 (1987).

7. C. J. Martin & A. N. R. Law, *Ultrasonics*, **21**, 85 (1983).

8. H. S. Fogler, F. G. Aerstin & K. D. Timmerhaus, 55th National Meeting, AIChE, Columbus, Ohio (1966).

9. (a) Kerry Ultrasonics, Hunting Gate, Wilbury Way, Hitchin, Herts SG4 0TQ, UK. (b) Lucas Dawe Ultrasonics, Concord Road, Western Avenue, London W3 0SD, UK. (c) ICI Chemicals and Polymers Ltd., Star House, 69–71 Clarendon Road, Watford WD1 1SQ, UK. (d) Embassy Machine and Tool Co., 104 High Street, London Colney, Herts AL2 1Q1, UK. (e) Durr Ltd, Broxell Close, Warwick CV54 5QF, UK. (f) Ultrasonic Power Services, Milton Works, Chapel Milton, Chapel-en-le-Frith, Stockport, Cheshire SK12 6QG, UK. (g) Maysonic Ultrasonics, Unit 10, Lonlas Village Works, Skewen, West Glam. SA10 6RP, UK.

10. (a) Kerry Ultrasonics, Hunting Gate, Wilbury Way, Hitchin, Herts SG4 0TQ, UK. (b) Maysonic Ultrasonics, Unit 10, Lonlas Village Works, Skewen, West Glam. SA10 6RP, UK.

Index

7DR